ROADSIDE GEOLOGY
of Idaho

David Alt
Donald W. Hyndman

MOUNTAIN PRESS PUBLISHING COMPANY
MISSOULA

Eighth Printing, September 2004

Library of Congress Cataloging-in-Publication Data

Alt, David D.
  Roadside geology of Idaho / David Alt, Donald W. Hyndman.
      p.   cm.
  ISBN 0-87842-219-6
  1. Geology—Idaho—Guide-books. 2. Idaho—Description and
travel—1981—Guide-books. I. Hyndman, Donald W.   II. Title.
QE103.A37 1989   89-36471
557.96—dc20   CIP

Mountain Press Publishing Company
P.O. Box 2399 / Missoula, MT 59806
406-728-1900

Preface

Idaho has some of the best rocks we have ever seen, and some of the loveliest and most interesting landscapes. Those rocks and landscapes are the raw material of geology, as well as the basic underpinning and natural setting for everything that lives in Idaho. Knowing them helps make sense of everything else. What is scenery if not rocks with trees growing on them?

We wrote this book for people who would like to know more about the geology of Idaho, but have neither the time nor the inclination to plow through the technical literature. Geology is not all that hard to understand if you can avoid most of the technical words. We hope we did that well enough to help our readers acquire a feel for the rocks and landscapes of Idaho.

This book starts with a chapter that briefly reviews the highlights of Idaho geology. The next four regional chapters each deal with the geology of a large part of the state: the Panhandle, central Idaho, the Snake River Plain, and the southeastern mountains. Each of those chapters starts with a more detailed review of the geology of their region, then continues with a series of roadguides that provide the local particulars. Between the chapters and the roadguides, we tried to convey some feeling for the rocks and landscapes, and some idea of what they mean.

Obviously, we didn't figure all this out ourselves. Most of the information in this book comes from the work of other geologists who have worked in Idaho since the early mining days. We did our best to distill the most interesting and most visible parts of their work into a more easily accessible and readable form. A good many of the geologists who have worked in Idaho will recognize themselves somewhere in this book. We want all of these colleagues of ours to know that we also recognize their contributions, thank them, and regret that the customs of publishing make it impossible to acknowledge each of their specific contributions in a book of this type.

This book also contains a few of our own ideas developed during our research on Idaho rocks. Those have already appeared in the technical literature. A long list of our graduate students participated in that research and helped formulate the new ideas it generated. We especially thank our colleague Jim Sears for helping create the theory about the big meteorite strike in southeastern Oregon and its role in starting the Snake River Plain and the Basin and Range.

The maps owe their attractive appearance to Dave Brower, who did the final art work on them. Dave Flaccus, Kathleen Ort, Jeannie Nuckolls, John Rimel, Rob Williams, and Kathy Spitler of Mountain Press all contributed in one way or another to setting type, designing the book, and laying out the pages. We thank all of them. We are also grateful to the highly skilled and professional editorial staff, who helped give the text its high literary polish that everyone so admires.

Finally, a plea. Some witty fellow once struck fairly close to the truth when he described geology as a vast web of theory tied together by a few road cuts and quarries, or words to that effect. In most areas, road cuts provide the best available exposures of the bedrock; they are very important to anyone interested in the Earth. Virtually all road cuts are interesting, and many expose rocks that are truly beautiful. Furthermore, they cost the taxpayers enormous amounts of money. Why should the highway department spend all that money to open an interesting exposure of beautiful rocks only to cover them with soil and then plant grass? We urge all our readers to oppose such thoughtless vandalism of costly and valuable road cuts. You can see grass almost anywhere, but it is hard to find a better place than a road cut to look at rocks.

And so we introduce you to our book on road cut appreciation in Idaho.

Dave Alt and Don Hyndman
Missoula, Montana
May, 1989

Table of Contents

ERA	PERIOD NAME AND AGE	IMPORTANT EVENTS IN IDAHO	
C E N O Z O I C	**Present**		
	Pleistocene	Latest ice age - ended about 10,000 years ago Earlier ice age - about 70,000 - 130,000 years ago Yellowstone volcano began to erupt about 1.8 - 0.6 million years ago At the end of the desert climate, modern streams begin to flow	
	——— 2.5 to 3 million years ago ———		
	T E R T I A R Y — **Pliocene**	Volcanic hotspot migrates northeast, leaving Snake River Plain in its wake	
	Miocene	A giant meteorite strikes southeastern Oregon, Basin and Range faulting begins, Columbia Plateau forms, Snake River Plain starts across southwestern Idaho	
	Oligocene	Time of wet and warm climate	
	Eocene	Shallow granites intrude and the Challis volcanic rocks erupt from them	
	65 million years ago — extinction of the dinosaurs		
M E S O Z O I C	**Cretaceous**	Idaho batholith and Kaniksu batholith intrude Rocky Mountains rise and overthrust belt stacks up to the east Oceanic islands dock against western Idaho to form Seven Devils and Riggins formations Thick sandstone, shale, gravel in eastern Idaho	
	Jurassic	Sandstones 2500 - 4500 feet thick in southeastern Idaho	Atlantic Ocean begins to open and a trench forms off the old west coast
	Triassic	Limestones and some shale 3000 - 4000 feet thick in southeastern Idaho	
	250 million years ago		
P A L E O Z O I C	**Permian and Pennsylvanian**	Sandstone, shale, and limestone 15,000 feet thick in southern Idaho, thinner farther east. Phosphate rock.	Sandstone and limestone 1000 - 2000 feet thick in central Idaho
	Mississippian	Dolomite and limestone 5000-10,000 feet thick in southern Idaho. Antler mountain building event.	Mudstone and sandstone 7500-10,000 feet thick in central Idaho
	Devonian and Silurian	Dolomite, limestone, and gypsum 2500 feet thick in southern Idaho, thinner farther east.	Dolomite 3500-6000 feet thick in central Idaho
	Ordovician	Limestone and some quartzite 3000 feet thick in southern Idaho, thinner farther east.	Dolomite to 1000 feet thick and sandstone to 3200 feet thick deposit in central Idaho
	Cambrian	Limestone, shale and sandstone up to 12,000 feet thick in southeastern Idaho.	Dolomite and sandstone deposit in central Idaho
	600 million years ago — first animal fossils		
P R E C A M B R I A N	**Proterozoic**	Old North American continent rifts to form new margin near western border of Idaho. Belt sandstone and mudstone accumulate to 50,000 feet thick in northern Idaho and southeastern Idaho. Basalt magma rises to inject sills in the lower part of the sediment pile. Older sediments metamorphosed to schists and gneisses	
	——— 2700 million years ago ———		
	Archean	Continental basement rock forms	
	4500 million years ago — formation of the earth		

MAP SYMBOLS

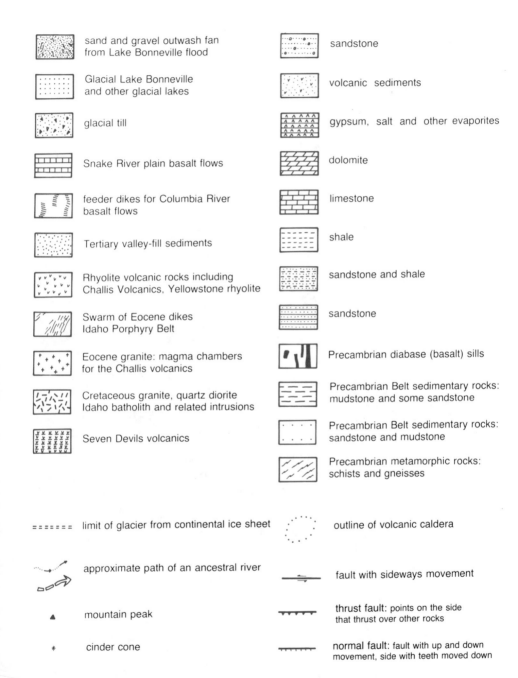

sand and gravel outwash fan from Lake Bonneville flood

sandstone

Glacial Lake Bonneville and other glacial lakes

volcanic sediments

glacial till

gypsum, salt and other evaporites

Snake River plain basalt flows

dolomite

feeder dikes for Columbia River basalt flows

limestone

Tertiary valley-fill sediments

shale

Rhyolite volcanic rocks including Challis Volcanics, Yellowstone rhyolite

sandstone and shale

Swarm of Eocene dikes Idaho Porphyry Belt

sandstone

Eocene granite: magma chambers for the Challis volcanics

Precambrian diabase (basalt) sills

Cretaceous granite, quartz diorite Idaho batholith and related intrusions

Precambrian Belt sedimentary rocks: mudstone and some sandstone

Seven Devils volcanics

Precambrian Belt sedimentary rocks: sandstone and mudstone

Precambrian metamorphic rocks: schists and gneisses

====== limit of glacier from continental ice sheet

outline of volcanic caldera

approximate path of an ancestral river

fault with sideways movement

▲ mountain peak

thrust fault: points on the side that thrust over other rocks

✶ cinder cone

normal fault: fault with up and down movement, side with teeth moved down

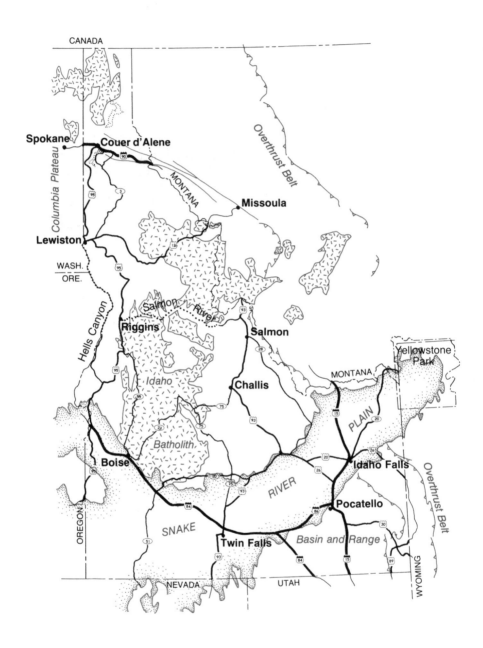

Roads and rivers covered in this book.

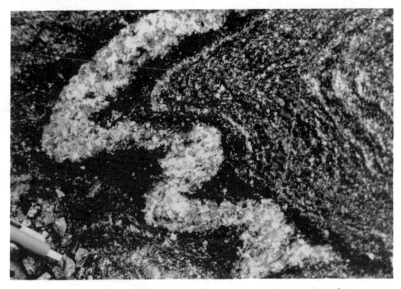

Contorted banding in this outcrop of gneiss gives some impression of the extreme pressure and temperature that prevailed as it formed.

I

The Big Picture

Continental Crust, Oceanic Crust

Idaho's oldest exposed rocks consist mostly of ancient schists and gneisses, metamorphic rocks that formed as still older rocks recrystallized at red heat deep below the surface. They also include large masses of granite that formed as molten magma crystallized, also far below the surface. Most of those rocks come in relatively pale shades of gray and pink. Almost all consist of crystals large enough that you can easily see them without a magnifier, and many are full of complexly swirling bands of light and dark. Most basement rocks really are quite beautiful.

Geologists call that complex of schist, gneiss, and granite the basement because nothing you can see at the surface suggests how deep those rocks may go. They lie beneath all the younger sedimentary and volcanic rocks that cover most parts of the continents. If you can summon the patience and financial

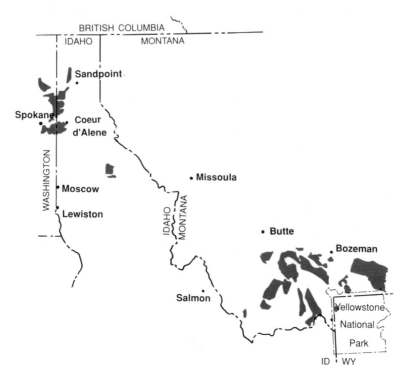

Basement rocks in Idaho and adjacent Montana.

forbearance to drill deep enough, you eventually reach basement rocks almost anywhere on a continent.

Basement rocks invariably look a mess. In truth, they are just as complex, just as baffling, as they look. Most were so thoroughly recrystallized, so profoundly metamorphosed, that they no longer resemble the original rocks from which they formed. It is usually extremely difficult, in many cases impossible, to discover what those original rocks may have been.

Neither is it easily possible to know the real ages of the basement rocks of Idaho. In much of the state, the relatively recent events that formed the Rocky Mountains heated the earth's crust enough to reset the radioactive clocks geologists use to determine the ages of rocks. That is why so many of the measured age dates on Idaho basement rocks are much younger than the probable age of the rock.

The very oldest age dates on Idaho basement rocks, therefore those least likely to have been reset, are in the neighborhood of 2700 million years. That is probably the time when they recrys-

2

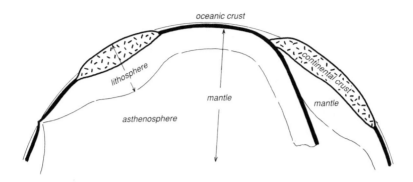

A cross-section of the earth. If this diagram were drawn strictly to scale, the continental crust would be about as thick in proportion to the planet as the skin of an apple is to the entire fruit.

tallized into metamorphic rocks; the original rocks may have existed for a long time before that. Another group of age dates clusters around 1500 million years. We think those dates probably tell of some later event that reset their internal clocks.

Basement rocks are the main stuff of the continental crust, a raft of relatively light rocks that floats on the much heavier rocks of the earth's interior, drifts along like an air mattress on a lake. The down and back time of earthquake waves echoing from the base of the continental crust shows that basement rocks probably extend to a depth of approximately 25 miles in most areas — no wonder that nothing we see at the surface gives a hint of how deep they may go. The oceanic crust is much thinner and utterly different.

There are places, some as close to Idaho as central Oregon, where big slabs of oceanic crust are exposed on land. You can

Continental and oceanic crust

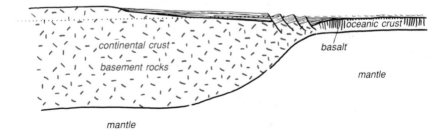

hike across oceanic crust in those places, see it exposed under the sky. Those exposures leave us in no doubt that the oceanic crust is about ten miles thick and consists almost entirely of black igneous rocks, basalt lava flows in the upper part, similar rocks that crystallized without erupting below. Basalt is distinctly heavier than continental basement rocks, so the oceanic crust floats lower than continental crust for the same reason that a half inflated air mattress floats lower than one that is full of air. Its lower buoyancy explains why most oceanic crust lies far below sea level.

Everywhere below either kind of crust lurks the mantle, the very bowels of the earth.

The Mantle

Mantle rocks comprise most of the earth, but you rarely see them at the surface. Western Idaho contains a few small areas of mantle rock, parts of western Oregon and Washington contain much more. Although they vary considerably, most mantle rocks are varieties of peridotite, black rocks quite noticeably heftier than the more familiar rocks of the continental crust. Mantle peridotites consist mostly of the common black mineral augite mixed with scattered grains of a pale green mineral called olivine.

In their native habitat within the earth, mantle rocks are so hot that they would glow bright red, if we could see them. In fact, rocks in the mantle are so hot they would melt if the miles of crustal rock above did not keep them under such enormous pressure. But solid though they are, those extremely hot rocks are so weak they flow plastically, like modelling clay. If something happens to reduce the pressure on them, or makes them slightly hotter, the hot rocks of the upper mantle partially melt to form basalt magma, the commonest volcanic rock.

PLATE TECTONICS

The group of theories that geologists call plate tectonics explains a great deal of what the earth does. Certainly they explain a great many of the events that shaped Idaho. As with many highly successful scientific theories, those of plate tec-

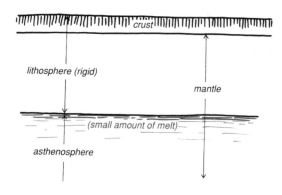

The lithosphere, a rigid outer rind on the earth, rides on the partially molten and slippery asthenosphere.

tonics are extremely simple. Anyone can understand them. They depend upon the basic idea that the earth has a more or less rigid outer rind called the lithosphere, which moves on a zone of slippery rocks beneath called the asthenosphere. The lithosphere consists of a dozen or so large pieces, called plates, which move independently. Plates may pull away from each other, collide with each other, or slide past each other. The moving plates create new ocean floor, destroy old ocean floor, and dismember or assemble continents.

Lithosphere and Asthenosphere

Although they may melt here and there, rocks in the upper mantle and crust are normally quite solid down to a depth of about 60 miles, even though capable of flowing plastically, like modelling clay. Geologists call that more or less rigid outer rind of the earth the lithosphere — unfortunately that long word has no vernacular equivalent. The lithosphere includes all of the earth's crust, and the outermost part of its mantle.

The solid lithosphere grades down into a deeper zone a hundred or so miles thick in which the temperature, which increases downward, is high enough to partially melt the rocks despite the pressure. That zone is the asthenosphere. The partially melted rocks of the asthenosphere are weak and slippery; they function as an internal banana peel beneath the solid lithosphere. At still greater depth, pressure again prevails over temperature to keep the rocks solid all the way down to the core of the earth.

The lithosphere covers the earth as a mosaic patchwork of a dozen or more large and small pieces called plates that fit together like the bones in a skull. Plates move constantly, apparently by sliding on the slippery asthenosphere. No overall pattern appears in the arrangement of the lithospheric plates, the directions of their motion, or their histories. So far as anyone can tell, they move randomly.

It may help to think of plates as big ice floes drifting on a slow current. Like the ice floes, plates can meet in any way. They may grind past each other, pull away from each other, or directly collide. All those possible kinds of plate motion played an important role in creating Idaho.

Oceanic Ridges — Separating Plates

Oceanic ridges, such as the mid-Atlantic Ridge, form where plates pull away from each other. Basalt magma rises through fractures that open between the separating plates and erupts into the widening gap between them to form oceanic crust. It is as though two ice flows were very slowly separating with water welling up into the gap between them, and then freezing onto their trailing edges. Both floes would grow larger as they separated, while the gap between them remained at constant width.

Basalt lava flows erupt at the oceanic ridge to form the upper part of the oceanic crust. Meanwhile, more basalt magma cools within the fractures to form vertical sheets of basalt, dikes, which form the middle part of the oceanic crust. Horizontal sheets of basalt magma that crystallize below the dikes form the lower oceanic crust, which lies on the black peridotites of the mantle.

An oceanic ridge. As the two plates separate at the crest of the ridge, basalt magma rises through the fractures and erupts between them to form new oceanic crust.

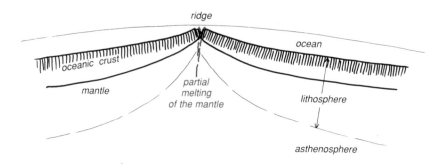

6

The situation becomes considerably more interesting if lithospheric plates begin to move away from each other where a continent happens to form the crust. Then, the separating plates split the continent, opening a rift between the pieces. As the plates continue to separate, the rift widens into a broad gulf, finally into a new ocean with the separated pieces of the continent on its opposite shores.

An oceanic ridge that split North America some 800 million years ago created a new west coast approximately along the western edge of Idaho. That coast existed until a plate collision finally destroyed it about 100 million years ago.

Oceanic Trenches — Colliding Plates

If the earth's oceanic ridges constantly create new oceanic crust, something else must constantly destroy oceanic crust. Otherwise, our planet would continually expand. The earth consumes its old oceanic crust in places where two plates collide.

A head on collision between two lithospheric plates is a horrendous thought, about as close as we can imagine to the mythical case of the irresistible force meeting an immovable object. But something must give; one of the colliding plates must yield. Invariably, the heavier of the two plates dives beneath the lighter, then sinks into the mantle.

A plate with oceanic crust on its surface slides through a trench and into the mantle, where it heats up and disappears.

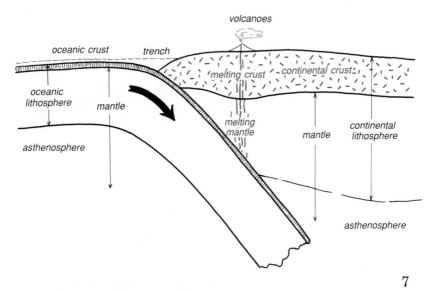

Without exception, the plate that sinks has oceanic crust on its surface; the light basement rocks of the continental crust can no more sink into the mantle than a marshmallow can sink into a cup of hot cocoa. If both colliding plates have oceanic crust on them, the one farthest from its generating ridge will sink because oceanic crust cools and becomes denser as it ages. The sinking plate absorbs heat as it descends, finally warming to the temperature of the surrounding mantle. As that happens, the sinking plate loses its identity as the relatively rigid outer rind of the Earth, and blends into the rest of the mantle.

An oceanic trench forms on the sea floor where the sinking oceanic crust turns downward to begin its long plunge into the mantle. By the time it meets its fate in the trench, most oceanic crust is at least tens of millions of years old, and has acquired a thick burden of muddy sediments, which are much too light to sink into the mantle with the heavy oceanic crust. So the sea floor sediments stay behind in the trench while the oceanic crust slides out from under them. They accumulate to great depth in trenches, tens of thousands of feet — a thick wedge of light material held low because it is resting on the descending oceanic crust.

Enough heat seeps in from the surrounding deep rocks to raise the lower parts of those trench filling wedges of sediment to a red heat. As that happens, the muddy sediments recrystallize to form gneisses and schists, melt to form granite magma. After the plate collision finally ends, no sinking plate holds the trench accumulation down. Then the whole complex mess of gneiss, schist, and granite rises to become a new piece of floating continental crust. That is how new continental crust

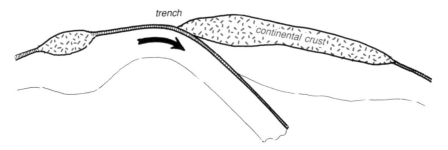

A trench reeling a small piece of continental crust onto a larger piece. When all the intervening oceanic crust is gone, the two pieces of continent will join.

forms, how the earth restores to the continents the crystalline basement rocks they lose to erosion. Trenches explain why, after several billion years of erosion and deposition, the continents still survive and the ocean basins are not brimful of sediment.

As trenches gobble up lithospheric plates with oceanic crust on them, they consume the sea floor that separates islands and continents, reel them into the trench at a rate of an inch or two a year. Any island or continent embedded in the sinking plate will eventually arrive at the trench. But most islands and all continents consist of rocks much too light to sink into the mantle, so they accumulate at the trench. That is how trenches assemble pieces of old continental crust into new continents.

Oceanic crust and the mantle rocks just below it consist in part of minerals that contain water. As the sinking plate heats, those minerals lose their water as they change into new minerals. The rejected water escapes from the sinking plate as red hot steam that rises into the already hot rocks of the mantle and crust. That steam reduces the temperature necessary to melt those rocks, forming large masses of magma. If rocks in the upper mantle melt, the magma has the composition of basalt, which generally erupts to form lava flows. Melting within the continental crust produces magma that will become granite if it crystallizes at depth; rhyolite if it erupts to become a volcanic rock.

A trench existed off the western edge of Idaho for something like 100 million years. During that time, it filled large parts of the state with granite, added groups of islands to its western boundary, and created the northern Rocky Mountains.

Transform Faults — Sliding Plates

Lithospheric plates simply slide past each other along transform faults. The notorious San Andreas fault of California is the most familiar example. Transform plate boundaries neither create nor destroy oceanic crust. But they do move large blocks of crust long distances, drastically rearranging the geography of continents and ocean basins. Such movements may well have swept pieces of western Idaho far north into British Columbia, but the details remain a bit obscure.

ANCIENT HISTORY OF IDAHO

Miles of covering rock must somehow disappear before the basement rocks that form in the red hot depths can finally see the light of day. If we assume that all those rocks were lost to erosion, that must have taken an inconceivably long time, probably hundreds of millions of years, at least. But those years left us without rocks to record their passing.

Whatever may have happened during the thousand or more million years after the basement rocks formed about 2700 million years ago, at least some of those rocks were exposed by about 1500 million years ago. That was when layers of sedimentary rocks that do survive began to accumulate on them. Those early sedimentary rocks piled up to overwhelming thickness, tens of thousands of feet in some areas.

Antique Sedimentary Formations

We speak here of sedimentary rocks that accumulated during Precambrian time, sometime during the four billion or more years that passed between formation of the oldest rocks and the beginning of Cambrian time. In all of science, few expressions are quite so simply encompassing as that term Precambrian.

Archean time is the earlier part of Precambrian time. It started with formation of the oldest rocks, whenever that may have been, and continued until about two billion years ago. That was the beginning of Proterozoic time, the latter part of Precambrian time, which lasted until the beginning of Cambrian time, about 570 million years ago. Sedimentary rocks accumulated in large parts of Idaho during much of Proterozoic time.

Throughout most of the Panhandle, as well as in large areas in the northern part of central Idaho, the Belt formations, Proterozoic sedimentary rocks, dominate the geologic landscape. They were named after the Belt Mountains of central Montana because that is where they were first studied. A name from western Montana or northern Idaho would have been far more appropriate.

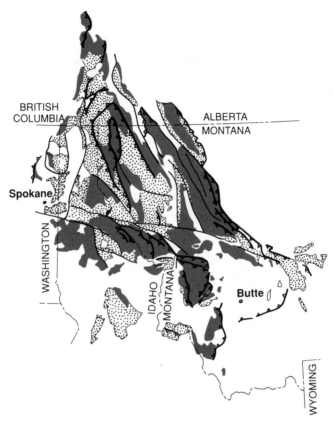

Distribution of late Precambrian Belt formations in Idaho and adjacent areas. Lower Belt stippled; upper Belt in color. —From Winston, 1988

The Belt formations of Idaho consist mostly of mudstones and sandstones in somber shades of gray and brown, along with some pale gray limestone. Central Idaho also contains the Lemhi and Yellowjacket groups of formations, which may actually be more Belt rocks. The mountains of southern Idaho contain still other Proterozoic formations, which will require a great deal more study before anyone can know how or whether they relate to the Belt rocks of central and northern Idaho.

It is hard to relate those different groups of sedimentary formations to each other because they exist in separate areas and contain no animal fossils that might provide internal evidence of their linkages. Age dates obtained by analyses of radioactive elements show that the oldest Belt sedimentary formations so far dated in Idaho were deposited sometime around 1500 million years ago, the youngest perhaps about 600 to 700 million years ago. The Lemhi and Yellowjacket formations are about the same age as the older Belt formations.

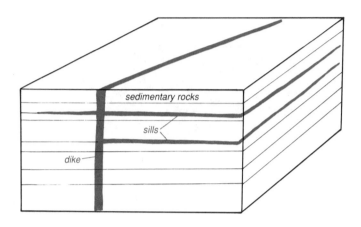

Sills are layers of igneous rock sandwiched between sedimentary layers. Dikes fill fractures.

All through northern Idaho and western Montana, the Belt sedimentary formations are full of big sills composed of diabase, a black igneous rock with the composition of ordinary basalt and a distinctive texture of its own. Sills of any kind form as molten magma squirts between layers of sedimentary rock to form a layer of igneous rock. Most of the sills are several hundred feet thick; at least one is more than a thousand feet thick. Age dates show that they were emplaced during a period that lasted from about 1500 to about 800 million years ago, the full length of time in which the Belt formations accumulated. That would seem incredible if the mechanical problem of injecting sills into a stack of sedimentary rocks did not lead to the same conclusion.

It is difficult to imagine how liquid magma could squirt between sedimentary layers and lift all those above to form a sill unless it were fairly close to the surface. The weight of a thick overlying accumulation of sedimentary layers would almost certainly make it impossible for sills to form. Therefore, the sills could not have been emplaced after the Belt sediments accumulated, but must have intruded them as they were laid down.

The Great Precambrian Continental Rift

Sometime late in Proterozoic time, perhaps about 800 million years ago, a rift broke the old North American continent, and the pieces drifted apart as a new ocean basin opened between them. Much of the evidence that tells of that event exists in northern Idaho.

The Belt Precambrian sedimentary formations of northern Idaho, all those tens of thousands of feet of them, end abruptly near the western border of the state. In the north, they extend a few miles into Washington; farther south, they don't quite make it to the border. Geologists find evidence that some of the Belt sediments washed in from the west, which is puzzling because no basement rocks old enough to have supplied that sediment now exist west of Idaho. Those who trace the Belt formations westward across Idaho find no hint that they are approaching the original western edge of the sediments. Instead, the Belt formations simply end as though they were chopped off, and so it seems.

Evidently, the plate broke into two pieces that drifted slowly apart, as a new ocean basin opened between them. The Windermere sedimentary formations of southern British Columbia and northeastern Washington were deposited about 800 million years ago. They look like the sort of stuff that accumulates along the edge of a continent, probably just as the opening rift was creating a new continental margin very close to the western border of Idaho.

The new western margin of North America that formed as the plate broke about 800 million years ago was to last for something like 700 million years. All that time, Idaho was on the west coast, and an open ocean stretched far beyond its western border. Remnants of the old continental margin and of the ocean beyond still exist along the western edge of the state.

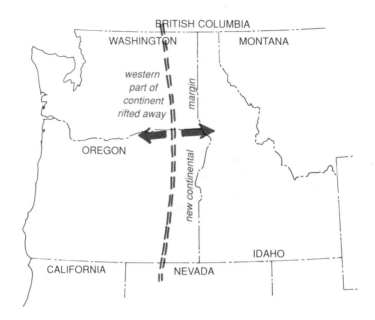

The new continental margin that formed during late Proterozoic time, about 800 million years ago.

Continental basement rocks, once formed, float on the mantle forever as a permanent part of the Earth's surface. Nothing can sink them, and erosion can't reduce them below sea level. So, the other piece of Precambrian North America, the part that contains the ancient rocks that eroded to supply sediment for the Belt sedimentary formations, must still exist, somewhere. Where?

Many geologists think the best candidate for the honor is a large piece of continental crust in Siberia. The basement rocks there are the right age, and the Precambrian sedimentary formations on top of them closely resemble the Belt and Windermere formations. Other geologists prefer to think that the missing piece of North America is now part of western Australia, another place where rocks of the right age and general appearance exist. Still others prefer some rocks in northern China and the northeastern Soviet Union. None of those ideas will be easy to prove until more geologists have an opportunity to study the rocks of both regions in question.

Paleozoic Sedimentary Formations

The beginning of Cambrian time is an important milestone in the Earth's history. That is when the oldest sedimentary rocks that contain abundant animal fossils were laid down. No one knows why animal fossils appear so suddenly in the Cambrian rocks with no geologic record of ancestors in Proterozoic rocks such as the Belt formations of Idaho.

The Paleozoic Era lasted from the beginning of Cambrian time, about 570 million years ago, until the end of Permian time, about 250 million years ago. During all that time, Idaho lay at the western margin of the North American continent, part of the western coastal plain.

Large parts of Idaho were submerged under shallow sea water during long intervals of Paleozoic time. Sediments that accumulated on the sea floor during those times added to the thick pile of Precambrian formations that already covered the basement rocks. Paleozoic sedimentary rocks exist in all parts of Idaho except the Snake River Plain. They are especially conspicuous in several of the mountain ranges in the southern part of central Idaho, and in those south and east of the Snake River Plain.

Paleozoic time ended at the end of the Permian period, about 250 million years ago. Something drastic happened then to destroy about 90 percent of all the animal species then living on the Earth. We think they died in the aftermath of a large meteorite impact in western Siberia. The north end of the Tungusska lava plateau marks the spot.

The Phantom Antler Mountains

Quite another kind of drastic event happened sometime around 300 million years ago. Geologists speak vaguely of the Antler mountain building event without knowing just what happened. Here and there along the old western margin of the continent, quite different sorts of rocks seem to record large crustal movements in quite different ways. Faults of that age exist in Nevada, granites in California, strange sequences of layered sedimentary rocks in central Idaho and northern British Columbia. It is difficult to learn what happened because formation of the Rocky Mountains destroyed or at least scrambled most of the record. All that remains is a few scraps, odds and ends of rock that whisper of long vanished mountains that preceded those we now see. Someday, when people have learned to listen with more knowing ears, those whispers may mean more.

So far, Idaho has yielded only scarce and enigmatic signs of those vanished Antler mountains; no evidence that would convince anyone if it did not fit into the pattern of the scattered signs that have turned up elsewhere. More revealing rock records of the Antler event will almost certainly emerge as work continues on the thousands of miles of rocks between southern California and the Yukon.

Mesozoic Sedimentary Formations

Mesozoic time began about 200 million years ago with the beginning of the Triassic Period. It continued through Jurassic time, and ended with the end of the Cretaceous Period about 65 million years ago. Those were the years of the Dinosaurs. At least some parts of Idaho still lay below sea level during those times, and more sedimentary formations accumulated.

All of Idaho's Mesozoic sedimentary formations are now rather sloppily stacked in the overthrust belt along the southeastern edge of the state, after having moved tens of miles east to get there. Those formations would cover a large part of southern Idaho if you could straighten them out and drag them back where they came from.

Central and panhandle Idaho contain no Mesozoic sedimentary formations, probably because none were laid down in those areas. The Rocky Mountains began to form during Mesozoic time, and much of Idaho may have been well above sea level before the Jurassic period had run its course.

The Rocky Mountain story started thousands of miles away in events that at first thought seem unrelated, certainly unlikely to cause such distant consequences.

The Big Plate Collision

Toward the end of Paleozoic time, the restlessly shifting plates assembled most of the Earth's inventory of continental crust into one great supercontinent, Pangaea. It didn't last

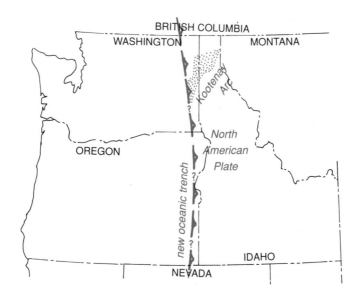

Westward movement of the North American plate broke the oceanic crust to establish a trench along the western edge of the continent.

long. Early in Triassic time, a bit less than 250 million years ago, the supercontinent began to break up along several rifts. The pieces are the modern continents we all memorized in grade school.

At first, the Atlantic Ocean was only a narrow gulf, like the modern Red Sea. It grew wider as the plates on opposite sides of the rift at its center drew away from each other. Meanwhile, lava flows erupting from that rift formed new oceanic crust between the separating plates, the floor of the Atlantic Ocean. If new oceanic crust forms in one part of the world, old oceanic crust must disappear at the same rate somewhere else. Otherwise, the Earth would steadily grow larger.

Westward movement of the North American plate away from the growing Atlantic Ocean jammed the western edge of the continent squarely against the floor of the Pacific Ocean. The plate broke along the western margin of the continent, and the Pacific plate began to slide down into the mantle as North America moved west across it. A deep oceanic trench formed on the sinking plate.

The Kootenay Arc

The first casualty of the plate collision was the sedimentary formations of the old continental shelf and coastal plain, which jammed together like the pleats of a collapsing accordion. Their tightly crumpled layers now make a fold belt called the Kootenay arc that trends south through southeastern British Columbia into northeastern Washington, then disappears beneath the younger basalt lava flows of the Columbia Plateau. It may well continue south along the western border of Idaho, but those sooty black flows spoil the view.

The Early Western Welt

The ocean floor sediments that rode the sinking plate into the depths of the trench were too light to follow it down into the mantle. Most scraped off into the trench, and now appear in places along the Kootenay arc. The sinking plate probably dragged some of the light sediments under the western edge of the continent, jacking it up in the same way that stuffing more logs under a raft will make it float higher.

Meanwhile, the force of the continuing plate collision compressed the western part of the continent, breaking it along fractures, and jamming the pieces over each other. Those fractures are called thrust faults, and movement along them thickened a broad zone along the western margin of the continent, raising its western edge still higher. The combined effects of telescoping the western part of the continent along thrust faults and stuffing oceanic sediments under it raised a broad marginal welt, a range of new mountains along the western part of the continent. That was the first stage in formation of the northern Rocky Mountains.

Meanwhile, red hot steam and molten basalt rising from the sinking plate into the already hot basement rocks above, melted large volumes of granite magma. It rose into the mountainous welt, making it so weak that it collapsed and slid eastward in great pieces to make the northern Rocky Mountains you now see.

The Overthrust Belt

Monstrously thick slabs of the upper part of the continent, pieces as much as ten miles thick, detached and slid off the still crystallizing magma in central Idaho, exposing the new granite almost as soon as it became solid rock. Those slabs moved east, rode thrust faults into western Montana. Neither the Panhandle nor southern Idaho contain as much granite as does the central part of the state, and the slabs that detached and slid east from those regions were not nearly so thick as those that rode out of central Idaho.

The enormously thick slabs that detached from the granite in central Idaho bulldozed the rocks ahead of them as they plowed east into Montana, raising big mountain ranges along their leading edges. The much thinner slabs that moved east out of the Panhandle separated into pieces that finally came to rest stacked and overlapped on each other almost like shingles on a roof. Those slabs now form the high mountains in the overthrust belt along the Rocky Mountain front in northern Montana. Relatively thin thrust slices that slid east out of southern Idaho stacked on each other in the mountainous overthrust belt of southeastern Idaho.

In all the areas that shed those slabs, you now see deep seated rocks exposed at the surface — continental basement, Precambrian sedimentary formations, and granite. Those deep rocks rose like an unloaded boat as their cover of younger rocks moved east. Remember that the continental crust is a raft of relatively light rock that floats on the mantle like a boat floats on water.

The average density of crustal rocks is approximately 80 percent that of mantle rocks, so the uplift after unloading will restore approximately 80 percent of the lost elevation. In other words, if a slab ten miles thick slid off part of central Idaho into Montana, the compensating uplift would amount to about eight miles, and the net loss in elevation would be about two miles. If that broad welt along the western margin of the continent had been 20,000 feet high before a slab 50,000 feet thick moved east out of central Idaho, it would have been about 10,000 feet high afterwards.

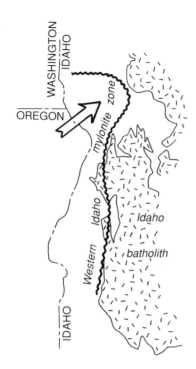

The western Idaho mylonite zone. The arrows show the direction of the lineation.

The Western Idaho Mylonite

A zone almost a mile thick of very strongly sheared rock, the western Idaho mylonite, traces long stretches of the old continental margin through the western part of central Idaho. It apparently formed just above the sinking plate, where its drag sheared the continental rocks overhead.

Mylonites are slabby rocks, good raw material for flag stones. If you look closely at the surface of a slab, turn it in the light to get the shadows just right, you can generally see a faint pattern of parallel lines, a lineation. The lines in the western mylonite all point northeast, apparently the exact direction of the shearing that created the mylonite, the direction of the plate collision. Evidently, the floor of the Pacific ocean was moving northeast as it slid under the old western margin of the continent. That direction is important.

20

Islands That Came into Port

Back at the end of Paleozoic time, while most of the Earth's continental crust was assembled into the supercontinent Pangaea, a few scraps still lurked unattached in the wastes of the Pacific when the plate collision began. They were miniature continents, like modern New Zealand, Borneo, or Japan. And there were chains of volcanic islands, like the Aleutians. As the trench along the old west coast gobbled Pacific Ocean floor at a rate of an inch or two a year, those scattered islands slowly closed on North America. About 100 million years ago, those vagrant scraps of crustal rock came ashore.

So far, their rocks have not revealed exactly where those islands wandered in from. Some of their sedimentary formations contain fossils more closely akin to those in eastern Asian than North American rocks of the same age. Those fossils are

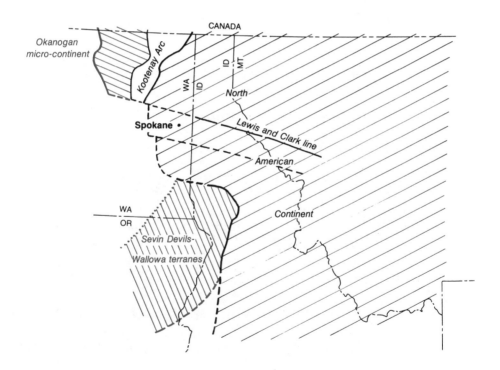

The islands that docked onto the western margin of North America approximately 100 million years ago.

related to modern animals that live in tropical waters. So, it seems reasonable to suppose that those islands may have come from some tropical latitude far to the southwest.

We might know how many wandering islands landed against western North America if that vast expanse of absolutely opaque basalt lava flows in the Columbia Plateau did not so completely obscure the older rocks. Nevertheless, there is no doubt that one large island became the Okanogan Highlands of northeastern Washington and southeastern British Columbia, a region so distinct that people recognized its geographic boundaries long before geologists realized that they also outline a miniature continent. The crushed remains of a large group of former oceanic islands now lies along the western edge of Idaho and in the Blue and Wallowa Mountains of Oregon, another highly distinctive area.

Many geologists speculate that the incoming islands pushed slabs of rock east into the overthrust belt as they docked. That idea is attractive and the timing is about right. Age dates on granites emplaced in faults associated with the overthrust belt generally fall in the range between 70 and 80 million years; dates on intrusions that solidified as the western Idaho mylonite zone stopped moving are about 80 to 85 million years. That zone marks the the suture between the former islands and the old continental margin.

The Trench Jumps West — Twice!

When those miniature continents and oceanic islands arrived at the old western margin of North America about 100 million years ago, they refused to sink like oceanic crust. Their arrival killed the trench, which was, after all, simply the place where the plate was sinking. But killing the trench didn't stop the plate collision because the plate bearing the North American continent was still moving west, an irresistible force. Something had to yield.

The plate bearing the floor of the Pacific Ocean broke again, this time west of the newly added islands. That created a new trench along a line that followed the Okanogan Valley of central British Columbia south into north-central Washington, then on south through Oregon.

22

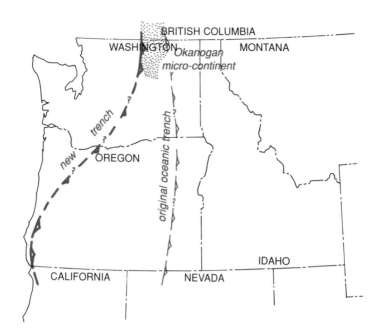

After the northern Cascade micro-continent joined North America about 50 million years ago, the trench jumped to its present position off the modern west coast.

That new trench swallowed oceanic floor until about 50 million years ago, when yet another small continent landed to form most of western Washington and British Columbia. Then the plate broke again, and the trench again jumped to a new line off the coast of Washington and Oregon, where it still swallows the floor of the Pacific Ocean.

Is Something Important Missing?

Many geologists argue that a large and important mass of rocks that should exist in western Idaho is somehow missing, and will feel that something very important is missing from this discussion if we don't consider the question. Specifically, they contend that the tightly folded Precambrian and Paleozoic formations of the Kootenay arc are not merely buried under younger lava flows, but are really missing. If so, their absence poses an interesting problem.

23

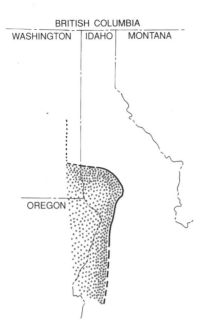

Is a big piece of rock mysteriously missing from this part of western Idaho?

Many of those who believe that the Kootenay arc is not where it should be in westernmost Idaho argue that the missing piece is in the Cassiar Mountains of northern British Columbia. They contend that northward movement of the Pacific Ocean floor swept those and other large blocks of rock hundreds of miles from their original homes. If so, that must have happened before the first wandering islands docked onto the old western margin of the continent.

Eocene Complications

Plenty of age dates on Idaho's igneous rocks fall in the range between 100 and 65 million years; another plenty fall close to 50 million years, Eocene time. Very few lie between. The second period of igneous activity happened long after the plate collision had shifted to the line through central Washington and Oregon. It is hard to imagine how such a distant collision could have caused such consequences in Idaho, especially since the igneous events of 50 million years ago extended east all the way into central Montana. The basic reason for that second great episode of igneous activity remains unclear.

The most obvious igneous activity of 50 million years ago was eruption of the Challis volcanic rocks. Enormous volcanoes erupted sheets of pale rhyolite ash that deeply blanketed much of central Idaho. Meanwhile, other large masses of magma crystallized beneath the surface to form numerous granite intrusions throughout much of central Idaho, as well as western and central Montana.

Western Cascades, Columbia Plateau, High Cascades

A new chain of volcanoes, the Western Cascades, rose along a line parallel to the trench, which by the end of Eocene time lay off the present west coast. All the evidence suggests that the Western Cascades were an extraordinarily active chain. Many of the eruptions must have been extremely violent. Red hot ash spread as much as a hundred miles east, well into central Oregon. Enormous volumes of airborne ash drifted east on the wind, then settled to fill the mountain valleys of the northern Rockies. Then, sometime around 17 million years ago, those impressive volcanoes of the Western Cascades abruptly snuffed out. As nearly as can be told, that happened just as volcanic activity was beginning in the Columbia Plateau.

The Columbia Plateau, mess of basalt flows in color.

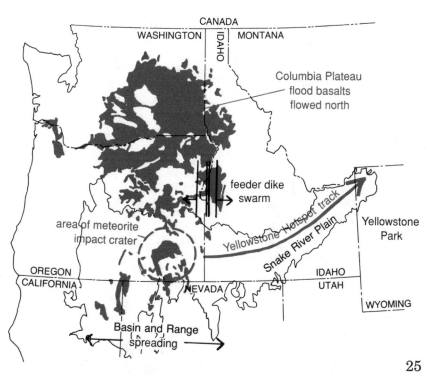

The Columbia Plateau, an enormous domain of basalt, covers most of eastern and central Washington and Oregon, the western fringe of Idaho, and a generous slice of northeastern California. Long ridges of the Blue and Wallowa mountains of Oregon divide the province into northern and southern parts. Most of the flows apparently erupted within a million or so years of late Miocene time, around 16 million years ago, in one of most intense episodes of volcanic activity known to have happened anywhere.

Their great size, more than anything else, distinguishes the lava flows of the Columbia Plateau. Geologists have traced individual flows all the way from eastern Washington to the Pacific Ocean, flows that travelled 300 miles from their source, covered many thousands of square miles, and contain dozens or even hundreds, of cubic miles of basalt. Those enormous floods of molten basalt must have spread within a matter of a few days

The High Cascades.

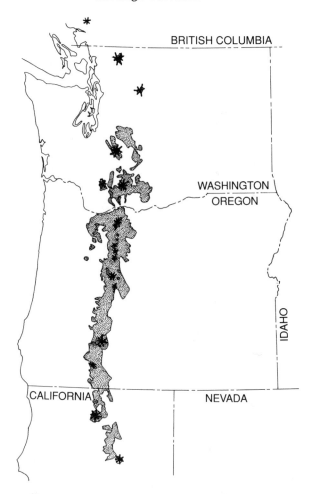

26

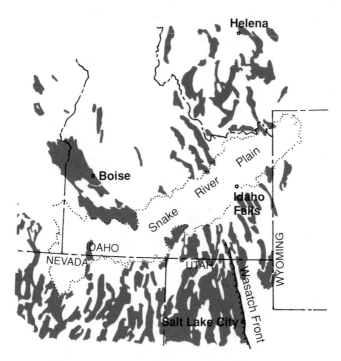

The northern part of the Basin and Range Province.

across areas that encompass large parts of a state. Then that spasm of intense volcanic activity stopped almost as abruptly as it had begun, and the Cascade volcanic chain woke from its long sleep.

A new row of volcanoes formed along a line distinctly east of the old Western Cascades, and began to build the long row of towering cones we see today. They depend for their supply of magma upon the ocean floor that continues to sink through the trench off the coast of the Pacific Northwest.

Basin and Range

One of the great geologists of the last century compared the mountains of the Basin and Range to an army of caterpillars marching north out of Mexico. And so they seem. With few exceptions, the ranges are elongated and aligned along a generally northerly axis. Each is an isolated block chopped out of the older Rocky Mountains, separated from the neighboring mountain blocks by broad expanses of flat basin floor.

Frequent earthquakes show that the faults that define the Basin and Range mountains and valleys are moving. Close study of those faults show that their movements are pulling the mountain ranges farther apart, making the valleys that separate them wider. The entire province is growing. It is tearing the continent apart.

As nearly as can be told, the Basin and Range province started developing 17 million years ago, almost exactly when the first basalt flows of the Columbia Plateau erupted, and just as volcanic activity ceased in the Western Cascades. Several million years later, a migrating volcanic hotspot began burning a broad swathe across southern Idaho, the Snake River Plain.

The Snake River Plain

At first glance, the Snake River Plain looks very much like a younger version of the Columbia Plateau, another nearly level volcanic plain. But the two provinces have very little in common. None of the flows in the Snake River Plain even remotely approaches the scale of the lava floods that built the Columbia Plateau. Furthermore, the basalt flows of the Snake River Plain are simply a thin carapace that covers vastly greater quantities of white rhyolite. Evidently, volcanic activity in the Snake River Plain began with eruption of enormous volumes of rhyolite, then concluded with a few basalt flows that spread a thin·crust of black lava over the rhyolite. The Yellowstone volcano illustrates the early stages in that process.

Most geologists regard the big volcano in Yellowstone Park, as the eastern extremity of the Snake River Plain. Similar but progressively older volcanoes appear to form a continuous chain that leads southwestward to Oregon, where the activity began about 17 million years ago. Many geologists interpret that chain of giant volcanoes as the track of a hotspot in the mantle. Evidently, the continent in its westward movement is crossing an abnormally hot part of the mantle that melts the rocks above it. A volcano forms above the hotspot, then goes out of business as the moving continent carries it past its source of heat. Meanwhile, a new volcano forms above the hotspot. The result is the long chain of volcanoes that becomes younger eastward as the continent moves west.

A CONSPIRACY OF EVENTS

The theories of plate tectonics seem not to reach far enough to explain this business of the Columbia Plateau with its associated events. No evidence suggests that anything peculiar happened at the trench off the west coast to cause the Columbia Plateau, the Basin and Range, or the Snake River Plain. Those events just happened, 17 million years ago.

What a curious choreography of geologic events that was. The Western Cascades snuffing out just as the Basin and Range and Columbia Plateau begin to develop. Then, a few million years later, volcanic activity winding down on the Columbia Plateau as the Snake River Plain and High Cascades began to develop. Can we dismiss all those congruences in time and space as merely a long series of casual happenstances, as meaningless? Or did something happen that might explain that sequence of coordinated late Miocene events?

One way to approach questions of geologic origins is to look for similar situations elsewhere, and then draw the necessary comparisons. The volcanic Columbia Plateau makes a good starting point. Our entire planet contains so few such regions that you can almost count them on your fingers: the Parana Plateau of South America, the Karroo Plateau of South Africa, the Deccan Plateau of India, and a few more. What do they have to say?

We will start with a discussion of the Deccan Plateau, about as far from Idaho as you can get. We start there because that is where we find the direct evidence that tells how lava plateaus and the things that go with them form — the smoking gun. Then, with the manner of the crime solved, we will return to the case in Idaho.

The Deccan Connection

The Deccan Plateau of western India is another great realm of basalt lava flows, overwhelming floods of basalt like those of the Columbia Plateau. The Deccan basalts also erupted within an amazingly short period, about 65 million years ago. There too, a number of other things began to happen in the same place at the same time.

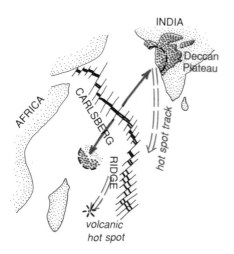

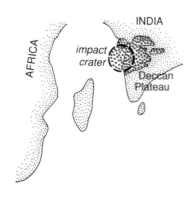

The Deccan Plateau of western India today, with its associated oceanic ridge and volcanic hotspot track.

The map with the ocean floor generated from the Carlsberg Ridge eliminated, to restore the situation of 65 million years ago.

The oldest ocean floor associated with the Carlsberg oceanic ridge, in the Arabian Sea between India and Africa, is 65 million years old. Evidently the Carlsberg ridge began to spread and create new ocean floor while the Deccan flood basalt flows were erupting. We think the Carlsberg Ridge is basically similar to the Basin and Range, the main difference being that it breaks oceanic instead of continental crust.

Meanwhile a volcanic hotspot began to create the long chain of the Chagos and Laccadive islands that extends south from the west coast of India. The oldest of those volcanic islands, at the northern end of the chain, is 65 million years old, and they become steadily younger southward. That southward progression reflects the northward movement of the Indian plate over a stationary hotspot in the mantle that was beneath western India when the Deccan basalts erupted. The Chagos-Laccadive hotspot track looks to us like the oceanic equivalent of the Snake River Plain.

If you cut the ocean floor generated at the Carlsberg ridge out of the map to collapse the picture to the situation of 65 million years ago, you bring the semi-circular arc of ocean floor that supports the Amirante Islands onto the coast of India. The Seychelles Islands, inside that arc, are a scrap of continental crust that evidently detached from India when the Carlsberg ridge formed. Spreading at that ridge has since moved both the Amirante and Seychelles islands far out in the Arabian Sea.

Obviously, something drastic happened on the western edge of India 65 million years ago to simultaneously create floods of basalt, a new oceanic ridge, and a new volcanic hotspot. The comparison with the events of late Miocene time in the Pacific Northwest is obvious. The differences between the two regions simply reflect the happenstance that the Columbia Plateau formed within a continent, whereas the Deccan Plateau formed at the edge of one. Horrible events that happened while the Deccan Plateau formed help tell the story of both regions.

The Terminal Cretaceous Calamity

The set of 65 million year coincidences goes far beyond western India. That was the occasion, the very geologic moment, of the great extinction that ended Cretaceous time. The Dinosaurs abruptly vanished then, in the very height of their glory, along with many other large groups of animals that lived on land and in the ocean. Does the great extinction somehow relate to the simultaneous events in western India?

Sedimentary rocks that were being deposited during the great extinction at the end of Cretaceous time contain a layer of black sediment an inch or so thick called the boundary clay. It cleanly separates rocks below that contain the fossils of Dinosaurs and the animals they knew from rocks above that are devoid of those fossils. The boundary clay has been closely studied in such widely separated parts of the world as Europe, New Zealand, and eastern Montana. It is about the same everywhere, a single thin layer of sediment that must have covered the entire planet.

The boundary clay is black because it is full of soot. Evidently enormous fires were burning as it was laid down. It also contains small mineral grains that show the distinctive effects of an extremely violent concussion, precisely the kind of intricate microscopic fracturing you can produce with dynamite, or a nuclear blast, but not with any lesser blow.

People who study fossil pollen see a record of the plant cover that covered the earth in time past. They find the sediments just below the boundary clay full of pollen from a wide variety of flowering plants, but the black layer itself and the rocks just above it contain very little pollen. Evidently, few plants bloomed while the boundary clay was laid down. That must indeed have been a time of terrible desolation, a catastrophe of widespread fires and of air filled with the dust that finally settled all over the Earth to become the boundary clay. To judge from the rate at which dust and soot settle, the sky must have been dark for at least several weeks, perhaps months. Then the air finally cleared and the sun again warmed the earth, but not in time to help the Dinosaurs and many other animals. They were already extinct, their bones bleaching in the returning sun.

Layers of sediment just above the boundary clay are full of fern spores; evidently the devastated landscapes of the time of the boundary clay greened first with ferns as the earth began to recover. Then pollen reappears in the following layers, recording the return of the flowering plants. The devastated earth was greening, probably within a year or two, but without most of its large animals.

All the varieties of pollen and other kinds of plant fossils reappear above the boundary clay in the southern hemisphere; a few plants never came back in the northern hemisphere. Some people interpret that strange pattern as evidence that the catastrophe struck in April, just when the northern hemisphere plants were emerging from winter hibernation and the southern hemisphere plants were going into dormancy. A catastrophe then would find the northern hemisphere plants at their most defenseless stage, those in the southern hemisphere in their least vulnerable season.

What happened?

Most geologists are now convinced that a large meteorite several miles across exploded as it struck the earth 65 million years ago. Only an enormous explosion can explain the shocked mineral grains in the boundary clay. Only the impact of a great meteorite at least several miles in diameter can cause an explosion so big and so violent. We suggest that the meteorite struck on the western edge of India, simultaneously creating the volcanic plateau, the hotspot, and the oceanic ridge. We think the mostly submerged semi-circle of the Amirante Arc is the western half of the crater rim, and believe the eastern half is buried under the Deccan Plateau. If so, then the crater is some 400 miles in diameter.

The crater was so deep that it relieved pressure on the hot rocks in the upper mantle enough to permit them to begin melting. Basalt magma then flooded the hole to make a lava lake, which overflowed like a giant soup bowl to send great flood basalt flows spilling across central India. The lava lake eventually cooled from the top down, meanwhile erupting great lava flows from the still molten interior across its congealing surface.

The exploding meteorite evidently split the rigid outer rind of the earth, the lithosphere, to start the Carlsberg Ridge, which has since opened the Arabian Sea with its continued spreading. The explosion started a crack that propagated into an oceanic ridge in about the way a flying rock sometimes starts a crack that works its way across a windshield. The opening crack also relieved pressure on the hot rocks in the upper part of the earth's mantle, permitting them to partially melt to form basalt magma. Eruption of that magma maintains the low pressure in the upper mantle, so partial melting continues for a very long time.

The hotspot started where the explosion blasted a mass of rock out of the upper mantle, permitting much hotter rock to rise from below. That deep wound remains stationary in its original place in the mantle as the lithospheric plate moves across it, generating the volcanic hotspot chain of the Chagos and Laccadive islands.

Back to Idaho.

The Columbia Plateau, Snake River Plain, and Basin and Range

If an exploding meteorite formed the Deccan Plateau as it started the Carlsberg Ridge and the hotspot that created the Chagos and Laccadive Islands, then another exploding meteorite must have formed the Columbia Plateau as it started the northern Basin and Range and the hotspot that created the Snake River Plain. The similarities between the two regions are so compelling that they must surely have a similar origin.

We think the meteorite that formed the Columbia Plateau struck about 17 million years ago because that is the age of the oldest basalt flows associated with both the Columbia Plateau and the Basin and Range. We think it struck in the southeastern corner of Oregon because both the hotspot track of the Snake River Plain and the northern end of the Basin and Range begin in that area. Large areas of volcanic rocks that erupted in southeastern Oregon about 16 to 17 million years ago are probably the remains of the lava lake.

Overflow from the lava lake repeatedly spilled northward through fissures in the Blue and Wallowa mountains into the lowlands of northern Oregon and Washington east of the Cascades. Those were the great floods of basalt that built the northern part of the Columbia Plateau. The same flows backed into the mountain valleys of western Idaho.

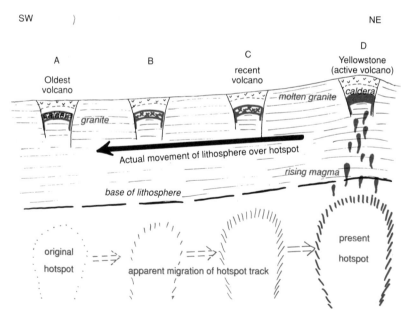

Continental lithosphere moving southwestward over a stationary mantle hotspot causes the hotspot track to migrate northeastward to its present position at the Yellowstone volcano.

Meanwhile, the Basin and Range began to propagate southward from southeastern Oregon, into central Nevada, perhaps extending as far as Mexico. Had North America been stationary, the Basin and Range would long since have split the continent along a line running south from southeastern Oregon; opened a new and growing ocean comparable to the Red Sea between the pieces. But this spreading fracture broke through a continent that was moving west, moving faster than its own spreading. So the Basin and Range is breaking the moving continent into blocks and pulling those apart, distributing its spreading over a broad region, instead of concentrating it along a single narrow line.

The eastern edge of the Basin and Range now lies along the Wasatch Front, the large fault scarp that defines the western face of the Wasatch Range in Utah. If you project the line of the Wasatch Front north, it passes right through the Yellowstone volcano. Evidently, the eastern edge of the Basin and Range and the Yellowstone hotspot are moving east at the same rate.

Of course, neither the hotspot nor the rift in the mantle under the continent, the Basin and Range, actually move. Both are fixed in position in the mantle, and only seem to move as the North American continent passes over them as it moves west. What will the future bring?

The Yellowstone volcano marks the present position of the hotspot deep in the earth's mantle. The hotspot left the Snake River Plain in its path.

If the present pattern of plate movement away from the mid-Atlantic ridge persists, the Yellowstone hotspot will continue to extend its volcanic track until partial melting finally stops in the mantle site of the impact. That will probably take many millions of years, probably long enough to propagate the Snake River Plain right through eastern Montana, possibly beyond. Meanwhile, the Basin and Range will also continue to migrate east at the same rate, leaving a shattered continent in its wake.

What if it Hadn't Happened?

The endless game of what if tends to be as futile as it is intriguing. What if John Wilkes Booth had missed when he took aim at Lincoln back in 1865? How would things have turned out if Lee Harvey Oswald's gun had misfired as John F. Kennedy rode through Dallas in 1963? What if that giant meteorite had not struck southeastern Oregon 17 million years ago? We can speculate.

As recently as 20 million years ago, volcanic ash flows erupting from the old Western Cascades spread eastward across a level plain. You can see them now in the Basin and Range, the same ash flow in one mountain block after another, showing none of the changes you might expect if they had moved across mountainous terrain. So, it seems that if the meteorite had missed, at least the northern part of the present Basin and Range would probably be fairly flat. The Rocky Mountains of central Idaho would probably continue south and west towards California as a region of modestly rugged hills. The isolated ranges of southern Idaho would not exist, nor would the broad valleys that separate them — Pocatello and Twin Falls would be anywhere from 60 to 100 miles closer than they are.

If the big meteorite had not struck southeastern Oregon 17 million years ago, the Yellowstone hotspot would not have burned its track through Idaho to create the Snake River Plain. This book would be one chapter shorter. The western High Plains from South Dakota to Texas would lack their deep deposits of volcanic ash erupted from the long chain of volcanoes that made the Snake River Plain.

Questions Waiting for Answers

We started this discussion of meteorites by mentioning a set of coincidences in time and space, in effect, a series of questions. We think the meteorite answers most of those questions, but some remain open. The most obvious of those is the matter of the Cascades. The timing certainly suggests that the meteorite strike somehow snuffed out the old Western Cascades, and that the modern High Cascades began to erupt as the region recovered from its effects. But the mechanism remains unclear. Why should a big hole in southeastern Oregon affect volcanic activity in the Cascades?

The warm and wet climate that accompanied eruption of the Columbia Plateau poses another fascinating question. Did the exploding meteorite affect the climate? Quite possibly.

Many people have suggested that enormous environmental effects far beyond mere dust and smoke followed the meteorite impact that created the Deccan Plateau at the end of Cretaceous time. The blazing passage of the meteorite through the atmosphere could have burned large amounts of nitrogen to create nitrogen oxides. Eruption of the flood basalt flows must have liberated large volumes of sulfur oxides, which would have combined with water in the atmosphere to form sulfurous and sulfuric acids. Those nitrogen and sulfur oxides in the atmosphere would then combine with water to form corrosively poisonous acid rain that could work havoc with living things on land and sea.

Why did the impact in southeastern Oregon not cause mass extinctions on the scale of those that followed the one in western India? Probably simply because the explosion in Oregon was not as big as the one in India. Possibly the season of the year was different. Perhaps it also matters if the meteorite strikes within a continent instead of on the edge of one. At this point, no one knows.

THE CHANGING CLIMATES
OF TERTIARY TIME

Tertiary time began about 65 million years ago with the sudden extinction of the dinosaurs and many other groups of animals, and still continues. Most geologists regard our own Pleistocene time as the most recent epoch of the Tertiary period. The record of changing Tertiary climates is clear enough to permit us to paint their picture, provided everyone understands that we use a broad brush with badly frayed bristles. That crude picture may shed some faint light on when the ancestors of the plants and animals that now inhabit Idaho first found the place congenial.

The Excessively Mild Weather of
Early Tertiary Time

Although Idaho is short of earliest Tertiary rocks, there are plenty that formed during Eocene time, which centered around 50 million years ago. We will start there, simply because that is where the best evidence starts.

Few forms of entombment are quite as sudden or drastic as being buried in a heavy fall of rhyolite ash. Central Idaho contains dozens of such time capsules in the Challis volcanic pile. Together, they preserve a rich record of Eocene conditions.

Here and there those deposits of Challis ash contain petrified redwood logs, even stumps standing where they grew. Modern redwoods grow in mild climates with plenty of moisture. If their ancestors of Eocene time preferred similar conditions, Idaho must then have enjoyed plenty of rainfall and mild winters.

The scarcity in Idaho of non-volcanic sedimentary formations laid down during Eocene time corroborates the climatic testimony of the fossil redwoods. Evidently, the climate was wet enough then to maintain a system of streams capable of carrying eroded sediment out of the state, probably all the way to the ocean. Very little stayed behind to become sedimentary rock formations. In the modern world, widespread deposition of sediment on dry land happens only in regions too dry to maintain a connected network of streams that drain to the ocean.

The First Long Dry Spell

Sometime around the end of Eocene time, about 40 million years ago, sediment began to accumulate throughout the Pacific

Northwest and northern Rocky Mountains. The John Day formation was laid down in central Oregon, the Renova formation in the broad valleys of the Northern Rockies, and sediments like those exposed in the Big Badlands of South Dakota spread across the northern High Plains. All those deposits resemble each other in being composed largely of silt, in containing rhyolite ash that apparently erupted in the Western Cascades, and in their abundant fossils of both plants and animals.

Whatever the rainfall may have been while all those sediments accumulated, it was not enough to maintain streams capable of hauling eroded sediment to the ocean. The sediments do contain petrified chunks of redwood and abundant impressions of redwood leaves, but no logs or stumps. Evidently, the big trees grew in the wetter high country; shedding leaves and pieces of wood into the dry valley floors far below. So the climate was mild enough and wet enough to permit redwoods to grow in the mountains, but not wet enough to maintain streams capable of carrying eroded sediment out of the valleys.

Sediment and volcanic ash continued to accumulate in the valleys throughout Oligocene time and well into the latter part of Miocene time, until about 17 million years ago, when conditions changed.

Tropical Idaho

About the time the enormous lava flows of the Columbia Plateau began to erupt, the climate of Idaho became very wet and very warm. Part of the record of that climate survives in deep red soils, laterites, like those that now form in the southeastern United States. More of the record exists in old lake beds that contain beautifully preserved fossil leaves, clearly the relics of subtropical hardwood forests. For a few million years, more or less, lush forests cloaked the mountains of Idaho.

Remnants of the red soils show up here and there as broad patches of red on the hillsides. They appear more predictably between lava flows in the Columbia Plateau, where they make ribbons of red soil sandwiched between much thicker flows of somber basalt. That association shows that the period of tropical climate coincided with the volcanic activity on the Columbia Plateau. Therefore, the numerous age dates on the basalt lava flows also tell us when the climate was so wet and warm.

Imagine those enormous lava flows quickly overwhelming tens of thousands of square miles of lush hardwood forest, trees disappearing under the molten basalt, then floating to the surface of the flow, and flaring like torches as they popped into the air. Burning forests undoubtedly filled the air with a dark pall of smoke that may have lasted for weeks as the fires raged. Meanwhile, the lava flows also created the lakes that preserved the leaves.

Wherever those enormous lava flows crossed rivers, they impounded lakes, which we can imagine becoming beautifully lush as soil began to form on the basalt and the regenerating forest reclaimed its territory. Leaves blown from the trees settled in the lake bottom muds, where occasional clouds of drifting volcanic ash covered them. No botanist ever preserved herbarium specimens more neatly, or nearly so permanently. Eventually, a new lava flow would fill the lake, securely burying those soft sediments with their hoards of pressed leaves.

Now, those buried lake sediments show up as white interbeds sandwiched between lava flows, the Latah formation of westernmost Idaho and eastern Washington. Layers of pale volcanic ash within the deposits split apart in thin slabs that show leaves starkly imprinted in black on their white surfaces. Many of the leaves are those of trees that live today in the southeastern United States and the Caribbean. It would be fun to know what exotic birds perched in the branches of those trees, what colorful butterflies flitted through their shade.

What a strange episode, that brief few million years of warm and wet weather. The curiously precise coincidence in timing suggests that the meteorite impact that created the Columbia Plateau may also have caused the episode of warm and wet climate. It is quite easy to imagine ways in which the two events could be linked, but very difficult to find solid evidence to prove or disprove those ideas.

Erupting volcanoes typically produce large volumes of carbon dioxide. Is it possible that the enormous eruptions in the

Columbia Plateau could have added enough carbon dioxide to the atmosphere to cause a greenhouse effect? If so, that might explain the warm climate, perhaps also the moisture. Erupting volcanoes also produce gaseous sulfur oxides, which combine with moisture in the air to become acid rain. Could eruptions on the extravagant scale of the Columbia Plateau have discharged enough sulfur oxides to cause genuinely serious acid rain for a long period of time?

Pliocene Sediment

Idaho dried up again long before late Miocene time ended, and remained dry through Pliocene time, until the ice ages began sometime between two and three million years ago. The evidence of that dry period exists in more sedimentary deposits, mostly gravel. Evidently, the big streams that drained Idaho during the wet interval of late Miocene time dried up as the rainfall became too meager to keep them flowing. If no streams carry sediment to the ocean, it accumulates on land as enduring evidence of the time the climate failed the streams.

Sediments of late Miocene and Pliocene age floor the broad valleys of the northern Rocky Mountains and the Basin and Range. And they lie sandwiched between volcanic rocks in the Snake River Plain. They include all the wide variety of sediments that accumulate in various parts of desert valleys.

The most typical late Miocene and Pliocene sediments are rather coarse gravels, commonly in some shade of brown. They probably accumulated mostly around the edges of valleys, in alluvial fans. Sand and silt travelled farther, and finally accumulated in the lower parts of the valleys. Because they are without stream drainage, many desert valleys contain lakes, which may be permanent or the kind that fill with water only during wet weather. In either case, the lake beds fill with thin layers of very fine sediment. All those sediments are alkaline, and therefore likely to contain fossil bones.

The late Miocene and Pliocene sediments contain very little fossil wood, perhaps because the climate was too dry then to support trees, even in the mountains. Animals that left their bones in the Pliocene sediments were generally similar to those that live in modern deserts — ancestors of our familiar camels, horses, rhinoceroses, and the like.

Valley-fill deposition ceased sometime between two and three million years ago, evidently because the climate became wet enough then to establish and maintain a network of streams capable of carrying eroded sediment to the ocean. That happened about when periodic ice ages began to afflict the Earth.

Ice Age Glaciation

Although they happened so recently that their mark still lies fresh on the land, no one really knows very much about the ice ages. Competing theories to explain them abound, but their cause remains unknown. Neither does anyone quite know what the ice age climates were like, or how many ice ages there were, or how long they lasted, or when the next one may start. It is frustrating to see so much evidence of the ice ages, and to understand so little of what it means.

Ice age counting has always been a dubious exercise. Each glacier scrapes, buries, and bulldozes the evidence of its predecessors, leaving intact only the record of earlier glaciers that extended beyond its own reach. For many years, most geologists thought that Pleistocene time saw four major ice ages come and go, with three long interglacial periods between them. Now, evidence from the deep ocean floors tells of a far more complex sequence of events still only dimly glimpsed.

Layers of mud accumulating on the sea floor pile up on their predecessors, instead of destroying them, and so retain a far more complete record than the brutally bulldozing glaciers left on the land. People who study deep sea cores have concluded that they can see evidence of many more than four ice ages — at least eight, perhaps as many as 20. No one can yet be sure.

Whatever the complete story may be, the landscape of Idaho contains a clear record of only two major ice ages, scattered and fragmentary evidence of earlier ones. We will discuss the two ice ages whose mark remains clearly etched on the land, and ignore the earlier glaciations whose record is no longer an obvious part of the landscape.

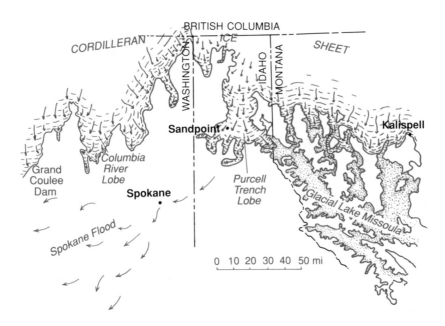

Maximum extent of the Cordilleran Ice Sheet. Glacial Lake Missoula and path of the Spokane Flood.

Glacial deposits of the earlier of the two well-recorded ice ages are too old to date by the radiocarbon method, which goes back only about 40 thousand years. They are too young to date easily by other radioactive clocks. That makes assigning an age more a matter of inspired guesswork than of quantitative science. For convenience, we will ignore all the controversy and use a rough guesswork figure of 100,000 years, hoping it is not too far from the truth.

Numerous radiocarbon dates leave little doubt that the most recent ice age reached its maximum about 15,000 years ago, and finally ended just 10,000 years ago. Those latest glaciers were not nearly as big as their predecessors of 100,000 or so years ago, so they left intact a large part of the record of the earlier glaciation. Glaciated regions of Idaho invariably contain moraines of the last glaciation nestled well within those of the previous ice age.

Ice creeping south out of British Columbia flooded the larger valleys of the panhandle during both of the major glaciations. Meanwhile, the high mountains of panhandle and central

Idaho supported valley glaciers. Glacial ice never filled the lower valleys of central Idaho, never lay on the Snake River Plain. Very few mountain glaciers gouged the high valleys of southern Idaho. Those areas emerged from the ice ages nearly unscathed.

The craggy high peaks and serrated mountain skylines of panhandle and central Idaho are the most obvious evidence of ice age glaciation. You see them in the higher mountain ranges, the ones that even today claw moisture out of the clouds to make their own weather. The jagged profiles of those ranges are the mark of the ice. Glaciers erode less as they begin to melt at lower elevations, where they leave their mark mainly in dumps of sediment. Ridges called moraines neatly outline the glacier that deposited them, preserving a precise record of its size and position. Muddy meltwater streams left smooth deposits of glacial outwash below most of the moraines.

The main effect in Idaho of the change in climate that started the ice ages was a general increase in precipitation. More water ended the second long dry spell that had begun in late Miocene time. Most of all, it started a new generation of streams.

The Modern Streams Begin to Flow

Widespread deposition of valley-fill sediments stopped about the same time the earliest ice ages began, sometime between two and three million years ago. That is the age of the youngest fossils in the valley-fill gravels. So it seems reasonable to suspect that the same climatic change that started the earliest ice ages also started streams flowing in the Northern Rockies. The new stream system, able to haul eroded sediment to the ocean, stopped deposition in the valleys.

In most parts of Idaho, especially the central and northern parts, the new streams just started flowing down old valleys that had not carried much water since the wet interval of late Miocene time. In some places, the streams had to establish entirely new courses. One of those was the Snake River Plain.

Much of the Snake River Plain formed during the dry climatic episode of latest Miocene and Pliocene time, so no old stream valleys were ready and waiting to carry the new drainage that began to flow as the ice ages began. The Snake

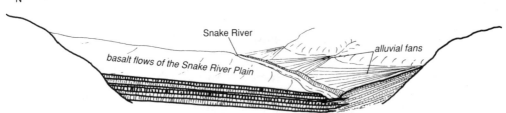

The Snake River began flowing along the lowest available course, where alluvial fans extending north lapped onto the south sloping surface of the Snake River Plain.

River and its main tributaries must have begun flowing along the lowest available paths, and then eroded those into new valleys.

The Snake River began to flow along a path close to the southern margin of the Snake River Plain simply because that entire broad surface has a gentle southward slope. But the southernmost edge of the Snake River Plain lay beneath alluvial fans built onto it from the mountains to the south. So the Snake River laid its course a few miles north of the southern edge of the Snake River Plain where the feather edges of the alluvial fans, sloping north, met the basalt surface, sloping south.

Southern Idaho was another place where the new streams had to find new courses in a situation far more complicated than that in the Snake River Plain. Any old valleys that may have existed in southern Idaho had been too chopped up by Basin and Range faulting to receive new streams. Furthermore, the dawn of the ice ages found that region a mosaic of undrained desert valleys like those that now exist in Nevada, eastern Oregon, and much of Utah.

The only way we can imagine to establish stream drainage out of an undrained valley is to fill it with water until the rising lake overflows through the lowest point on the drainage divide. Then, the flow through the spillway will eventually erode it deep enough to drain the lake. We believe that most of the stream drainage of southern Idaho started flowing in just that way after the climate turned wetter as the ice ages began. A large part of that drainage goes to the Great Salt Lake, which still has no outlet.

The idea of starting a stream drainage by filling an undrained desert basin until it overflows, then waiting for the spillway to drain the lake is not even slightly speculative. That is exactly what happened in southern Idaho about 15,000 years ago, under the wet climate of the last ice age. Consider the story of Lake Bonneville.

The northern part of Lake Bonneville.

During the last ice age, and presumably the earlier one, the Great Salt Lake swelled into ancient Lake Bonneville, which covered a large area of northern Utah, and flooded well into southern Idaho. The lake overflowed through Red Rock Pass in the hills southeast of Pocatello, then down Marsh Creek and the Portneuf River to the Snake River. If the wet ice age climate had continued, that overflow stream would still be flowing down that course, gradually entrenching its channel, eventually to drain the Great Salt Lake.

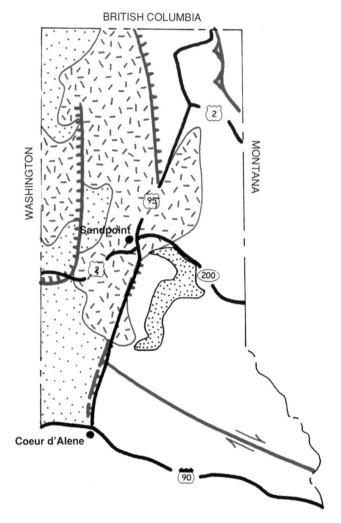

BRITISH COLUMBIA

WASHINGTON

MONTANA

Sandpoint

Coeur d'Alene

Roads covered in this section.

II
The Panhandle

If we were to attempt to describe its bedrock in the fewest possible words, we would call northern Idaho a region of Precambrian sedimentary formations. By a generous measure, the most abundant rocks there are the Belt formations that were laid down during Precambrian time, which ended about 570 million years ago. Most of those formations are more than a billion years old — quite an impressive age, even for rocks.

No one really knows how much Precambrian sedimentary rock exists in northern Idaho. The original sediments were laid down long before animals lived on the earth, so they contain no fossils that might provide internal evidence of their age. That makes it extremely difficult to match layers of rock from one place to another, almost impossible to measure the thicknesses of the formations. The best estimates place the minimum total thickness of all the Precambrian sedimentary formations at several tens of thousands of feet. Those figures, vague as they are, would seem ridiculously large, absurdly unrealistic, if the evidence in their favor were not so powerfully convincing.

The Belt Basin

It is hard, virtually impossible, to imagine how tens of thousands of feet of sediment could possibly have accumulated on top of the continental raft of light rocks. The Belt sedimentary formations must fill a hole of some sort, a very deep sedimentary basin. How did that hole form?

Most geologists who have thought deeply about the Belt sedimentary rocks have concluded that they accumulated in a narrow ocean basin, perhaps one that was slowly growing wider. That could explain most of the great thickness, but not why the Belt sediments accumulated in the limited area of the northern Rockies, instead of in the much longer, much more linear, deposit that you might expect to form along a coast. A few geologists, ourselves among them, toy with the idea that the Belt basin may have begun its existence as an enormous meteorite explosion crater several hundred miles across and tens of thousands of feet deep. At the moment, no one can produce compelling evidence to favor either point of view.

Diabase Sills

Whatever its origin, the Belt basin filled with igneous as well as with sedimentary rocks. The Belt sedimentary rocks of northern Idaho contain numerous large diabase sills, layers of black igneous rock sandwiched between the sedimentary beds. Most of the sills are several hundred feet thick, a few more than a thousand feet. It seems likely that many or most of the sills extend over very large areas, but the present status of geologic mapping in the northern Rocky Mountains does not definitely demonstrate that they do. Many other sedimentary basins of approximately the same age contain similar swarms of diabase sills, some of which certainly do extend over areas of many thousands of square miles.

Age dates on the diabase sills in the Belt formations range from about 1500 to about 800 million years. Most of the sills are somewhat altered, so their radioactive clocks may have been reset to ages that are too young. It is also possible that the dates are accurate, that the sills really were injected over a period of some 600 or so million years. Age dates on similar swarms of sills in several other large sedimentary basins cover a similar range.

In fact, the mechanical problem of squirting a layer of magma between layers of sediments does make it seem reasonable to argue that the sills really were injected over a long period. To form a sill, the magma must lift all the overlying layers of sediments. That makes it much easier to imagine a sill forming near the top of a sequence of sedimentary layers than

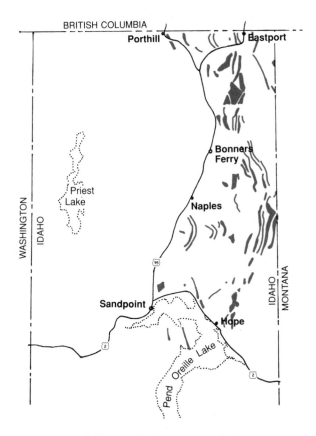

The largest diabase sills of northern Idaho. Many others are too small to show on this map.

near the base simply because it would not need to lift such a heavy burden. The age dates on sills in the younger formations are consistently younger than those in the older formations, which suggests that they were indeed injected near the top of the accumulating section.

Many geologists interpret that great swarm of sills as a record of the time when a big western chunk of the old North American continent rifted off; this probably happened sometime around 800 million years ago. The Precambrian Windermere formations of British Columbia and northeastern Washington are about 800 million years old, and look like the kind of stuff that would accumulate in an opening rift. So, about 800 million years ago may be when a big piece of North America detached and drifted off, when ocean waves first began to break along a new west coast that lay very close to the present western border of Idaho.

If the continent split 800 million years ago, then what are all of those older sills doing in the Belt formations? In fact, quite a number of large sedimentary accumulations that fill deep basins contain swarms of diabase sills, and some of those are in no way associated with a continental rift. Great swarms of diabase sills appear to be typical of such basins, by no means a peculiarity necessarily associated with rifting.

All these thoughts make it seem a pity that so little is known of the diabase sills in the Belt formations. Most geologists have assumed that they have little or nothing to do with the sedimentary formations that contain them. We think they are an integral part of those formations, a vital piece of the puzzle.

The Vanished Precambrian World

Rocks of any kind, sedimentary rocks in particular, open a window into the past by preserving a record of the conditions in which they formed. The Precambrian sedimentary rocks of northern Idaho give us a narrow glimpse through dimmed and discolored glass into a strange and unknown planet. They show us an Earth so strangely different from our cozily familiar world that it could be a planet as distant in space as it is in time.

Superficially, those ancient rocks look like ordinary sandstones, mudstones, and limestones. But that friendly look probably reflects our human tendency to seek something familiar in everything we see, especially in strange things. In fact, a longer and more thoughtful look at the Belt formations reveals all sorts of peculiarities.

For example, Belt sedimentary rocks contain much less clay in proportion to silt and sand than do most younger rocks. Did minerals not weather into clay a billion years ago? Evidently not. The silt and sand show little or no evidence of having moved before the wind. Did no wind blow a billion years ago? If so, the rocks contain no evidence of it. Belt rocks contain no animal fossils. Were there no animals a billion years ago? Apparently not.

Exquisitely preserved ripple marks on a slab of Precambrian sandstone. These are the kind of ripples that form under water.

That distant planet on which those strange rocks accumulated is forever vanished, evolved into a completely different world. We will never see it, never know for sure what it was like. But the rocks are not really quite as mute as the proverbial stone. They whisper suggestions that can inspire reasonable conjecture.

It is easy, for example, to imagine why the Belt rocks contain so little clay. The loftiest plants that grew then were blue-green algae, the same scum that grows in some roadside puddles today. They could not have spread an umbrella of leaves over the ground to shelter it from splashing rain drops. Modern deserts where very few leaves shelter the soil from rainsplash typically have a high rate of erosion and a scanty soil cover. It isn't hard to imagine the Precambrian rains stripping the rocky surface of the landscape so efficiently that very few soils lasted long enough to break down into clay minerals.

What about the wind, or perhaps the absence of wind? The Belt formations include thick accumulations of sandstone and siltstone that certainly accumulated on land. Yet there are no sand dunes, no wind ripples, no concrete evidence that the wind ever blew hard enough to move sand or silt. The little wave-shaped sand ripples on some bedding surfaces may be evidence of gentle breezes that ruffled the water surface without moving much sand or silt.

There is reason to believe that the late Precambrian atmosphere contained very little oxygen, reason to imagine that it probably contained large amounts of carbon dioxide. Carbon dioxide blocks heat radiation in much the same way that glass does to create a greenhouse effect, a situation in which sunlight can enter, but heat radiation can not escape. So an atmosphere rich in carbon dioxide would keep the earth very warm. And that would raise the evaporation rate from the oceans enough to make the atmosphere extremely humid as well as very warm. For the sake of argument, imagine the Precambrian world as a sort of planetary sauna bath.

Water vapor holds and conducts heat more efficiently than other atmospheric gases. An atmosphere rich in water vapor might enable the earth to even out temperature differences by conducting heat from one place to another, instead of relying on wind for heat transfer. Furthermore, an atmosphere very rich in water vapor would probably maintain a continuous cover of clouds, which would favor more uniform solar heating of the earth's surface, thus minimizing the need for heat transfer from one region to another.

Nearly everywhere Precambrian sedimentary formations exist, they include thick sections of rocks that were deposited on land. Why did all that sediment linger on the land, instead of going down rivers to the sea? Perhaps carbon dioxide in the atmosphere kept the earth's surface hot enough that streams tended to evaporate before they could reach the sea, leaving their burden of sediment on the land.

Whatever its composition, the earth's Precambrian atmosphere was not hot enough to boil water. Many of the Belt sedimentary rocks contain abundant evidence of having been deposited in surface water. Furthermore, those rocks are full of fossil blue-green algae, which still thrive in such varied places as roadside puddles, the north sides of damp tree trunks, the Great Salt Lake, and the hot springs in Yellowstone Park. But blue-green algae can't survive in boiling water, so the Precambrian climate could not have been that hot.

To judge from the abundance of their fossil remains, the blue-green algae must have thrived mightily in Precambrian time, probably spreading a scummy growth wherever they could find water and light. Their modern descendants, which look just like the fossil remains of their Precambrian ancestors, also live almost anywhere. Give their scummy green films the respect due the descendants of the plants that made the earth into a place fit for us animals.

Like all green plants, blue-green algae absorb carbon dioxide from the atmosphere, use the energy of sunlight to extract the carbon from it to make their tissues, and release free oxygen to the atmosphere. If accumulating sediments bury the plant after it dies, the carbon will be trapped in the sediment, and the oxygen that was once attached to it will remain in the atmosphere. Trapped carbon compounds that stain many of the Precambrian sedimentary formations dark brown or black show that oxygen was certainly accumulating in the atmosphere as the sediments were laid down — one molecule of oxygen added to the atmosphere for every atom of carbon incorporated in the sediment.

As plants growing under water withdraw carbon dioxide from the water near them, they make the water slightly alkaline, and that causes calcium carbonate to precipitate —

Fossil blue-green algae in Precambrian sedimentary rocks.

more carbon dioxide taken out of circulation. Calcium carbonate sediment solidifies into limestone. Every little bit of limestone is a little bit of carbon dioxide taken from the atmosphere and crystallized into the earth's crust. The Precambrian sedimentary formations contain a great deal of limestone; their deposition must have reduced the amount of carbon dioxide in the atmosphere, made the earth cooler.

As Precambrian time drew to a close, the blue-green algae had finally extracted enough carbon dioxide from the atmosphere and added enough oxygen to the atmosphere to make it possible for primitive animals to breathe. The earliest animal fossils appear as faint impressions in the very latest Precambrian sedimentary rocks. Then suddenly, about 570 million years ago, all sorts of animals spread through the oceans. That event marks the end of Precambrian time, the beginning of Cambrian time. The earth hasn't been the same since.

Ever since animals first appeared in abundance, they have squirmed, dug, and eaten their way through any muddy sediment they could find. All that activity stirs the sediment; destroys fine structures such as ripple marks, mud cracks, and raindrop imprints. The Precambrian sedimentary rocks retain all that fine detail, every kind of structure and mark we commonly see in soft sediments — but no footprints, no sign of animals. Watch for all the little sedimentary details so elegantly preserved in the Precambrian sedimentary formations of northern Idaho. They commonly show up in stream worn pebbles as swirling patterns of fine lines that give the rock a distinctively Precambrian look.

They hardly look it, but all of the Belt formations of northern Idaho are somewhat metamorphosed, recrystallized at high temperature. Most metamorphic rocks that recrystallized at the same temperature as the Belt formations are slates, which split into thin slabs along fractures that cut across the original sedimentary layers. Few of the Belt rocks show such slaty cleavage. Some people call them argillites, a term often applied to very hard and somewhat recrystallized mudstones that are not slates.

Slates develop from rocks that recrystallized while they were being crumpled into folds. Evidently, the Belt rocks simply recrystallized with no accompanying deformation. The extremely deep pile of sediments with all those sills of molten diabase in them got hot enough to bake the rocks. The Belt rocks of northern Idaho acquired their folds long after they recrystallized, as the Rocky Mountains formed during late Mesozoic time, about 70 or 80 million years ago.

The Kaniksu Batholith and the Purcell Trench

During Cretaceous time, about 70 to 80 million years ago, large masses of granite magma rose into the upper part of the earth's crust in northern Idaho. That was about the time the great granite batholiths of central Idaho formed, and the geologic setting was the same. The main difference was that the volume of magma was less in northern than in central Idaho.

Even so, it seems that the volume of granite magma was enough to weaken the earth's crust, permitting its upper part to shear off and move east. Now you see that detached slab in the Purcell and Cabinet mountains along the eastern side of the panhandle, the granite in the Kaniksu batholith of the

The Purcell and Cabinet mountains consist mostly of Belt formations that covered the granite of the Kaniksu batholith before they moved east.

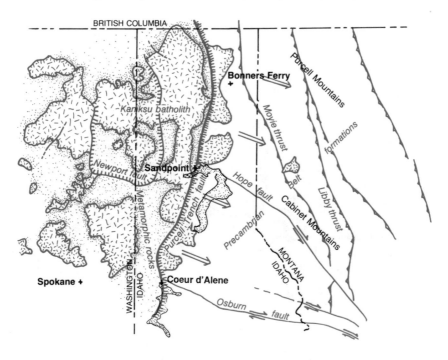

Selkirk Mountains along its western side. The Purcell trench separates them. Several fairly small masses of granite in the Purcell and Cabinet mountains appear to be parts of the Kaniksu batholith sheared off and carried east.

All that makes the Purcell trench one of the most important structural elements in northern Idaho. It is a pity that deep deposits of glacial debris and older sediments in its floor make it so difficult to find enough bedrock exposure to reveal more about the structure. Nevertheless, it is clear that throughout its great length the Purcell trench separates two geologic provinces. Mountains on its west side consist of rocks formed deep in the earth's crust; those on its east side contain rocks formed at much shallower depth.

Here and there, the easternmost surface of the mountain front west of the Purcell trench is mylonite, a slabby metamorphic rock with a peculiar texture that speaks of much shearing. We have actually seen faint remnants of that mylonite only south of Sandpoint, but assume that it probably continues north under a cover of glacial deposits. The simplest interpretation has the mylonite forming in a fault zone that penetrated deeply enough in the earth's crust to find rocks hot enough to recrystallize as they sheared.

Section showing the Purcell trench interpreted as the result of movement along a gently dipping fault that is slightly concave upward.

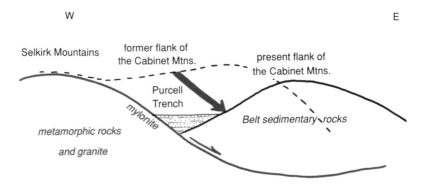

Imagine it: the mountain ranges east of the Purcell trench shearing the mylonite as they moved east off the mountains to the west. Imagine the fault as a slightly curved surface, concave upward, and the rocks moving east are left with no alternative except to rotate down to the west, thus creating the depression we know as the Purcell trench.

The timing of the crustal movements that created the Purcell trench remains unclear. One plausible scenario has the rocks east of the trench moving during late Cretaceous time, about 70 million years ago, while the magma that became the Kaniksu batholith was still partly molten. Another interpretation, also plausible, places the movement in Eocene time, about 50 million years ago. That was a period when the earth's crust was being stretched throughout much of western North America, and the Purcell trench could fit comfortably into the picture. Resolution of the question awaits better methods of dating mylonites.

The Coeur d'Alene Ores

One of the problems in the Coeur d'Alene district is that the ores appear to be older than the veins that contain them. Lead isotope studies seem to show that the metals in the ores separated into the rocks about 1.2 billion years old, while the Belt formations were accumulating. But the ore veins are clearly associated with the Lewis and Clark fault system that formed with the Rocky Mountains, probably between 70 and 90 million years ago, possibly as recently as 50 million years ago. What could that mean?

It now seems likely that the two dates refer to different events: the first put the metals into the rocks, the second concentrated them into veins. So we may have the rare and happy situation in which both sides to an old argument turn out to own a piece of the truth. First, let's put the metals into the rocks.

Assume for the sake of discussion that the North American continent first began to split sometime around 1100 million years ago. If so, a new ocean basin opened then along the line of the rift as the two pieces of continent separated. An oceanic ridge similar to the one that runs down the middle of the modern Atlantic Ocean would have existed off the west coast of Idaho. It could have been responsible for the original deposition of metals in the rocks.

Research submarines exploring oceanic ridges have revealed enormous hot springs on the ocean floor. Blistering hot water rises from cracks in the ocean floor, looking like great columns of white or black smoke billowing from factory smokestacks, white steamers and black smokers. The smoke consists largely of metal sulfides that precipitate from the hot water and settle on the ocean floor, just as factory smokestacks drop fly ash and soot on everything beneath their plume. Those submarine mineral deposits accumulate to form sedimentary rocks that contain layer upon glittering layer of metal sulfide minerals.

That may be the how the Belt sedimentary formations acquired their content of heavy metals about 1.2 billion years ago. The metals concentrated in the western part of the Belt sedimentary sequence because that part was near the new oceanic ridge.

Then, about 70 million years ago, movement on faults of the Lewis and Clark zone opened fractures in the ancient Belt formations. Minerals, including metallic ore minerals, moved out of the sedimentary rocks and into the open fractures to form the vein deposits that the mines work. So the veins that formed about 70 million years ago contain metals that separated from sea water some 1.2 billion years ago.

Glaciation

The last decade or two has seen the question of how many ice ages there were and when they happened fill with treacherous uncertainties. This is not the place, nor are we the authors, to resolve the nasty issues. They need not bother us here.

Despite all the controversy, it seems evident that only two ice ages left a clear record in the landscape of the northern Rocky Mountains. The earlier happened at an unknown time, perhaps around 100,000 years ago. The most recent reached its climax about 15,000 years ago, its final end sometime around 10,000 years ago. The record of other ice ages that may have happened long before 100,000 years ago has nearly disappeared from the landscape, but survives in occasional patches of glacial sediment.

During both of those ice ages, glaciers that covered most of British Columbia flowed south into northernmost Idaho, filling the valleys and covering all but the higher mountains. Those higher mountains supported glaciers of their own that flowed down the valleys to join the regional ice. In the broad valleys of northern Idaho, the main mementos of both ice ages exist in widespread deposits of glacial debris — till and outwash.

Glacial till is sediment dumped directly from the ice, an unsorted and unlayered mixture of all sizes of sediment from clay to boulders — unsightly stuff. Any deposit of till is a moraine, direct evidence that glacial ice once existed in exactly that spot. The most conspicuous and informative deposits of till are the morainal ridges that precisely outline the former boundaries of the glacier.

Outwash is deposited from glacial meltwater; it consists of neat layers of clay, sand, and gravel. A smooth blanket of outwash extends downslope from most moraines, a remembrance of the torrents of muddy meltwater that swept across those slopes during ice age summers. Since the last ice age ended, streams have entrenched themselves into most of those outwash deposits, leaving remnant benches along the sides of the valleys.

Valley glaciers formed on the higher mountains that stood above the level of the regional ice. Their impression on the landscape consists mostly of erosional rather than depositional landforms: deeply gouged mountain valleys that lead upward into deep cirques hollowed below craggy peaks. Lakes sparkling in rock rimmed basins in the high mountain valleys and brightly tumbling waterfalls owe their existence to the tendency of glaciers to erode the valley floor unevenly.

Glacial Lake Missoula

Regional ice of both major ice ages flowed south down the Purcell Valley far enough to dam the Clark Fork River at the present site of Pend Oreille Lake, impounding Glacial Lake Missoula. At its maximum filling, that lake was about 2000 feet deep at the ice dam, and flooded the Clark Fork drainage of

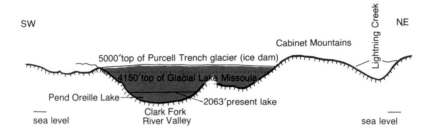

Glacial Lake Missoula was dammed at the east end of Pend Oreille Lake by a large lobe of glacial ice that filled the Purcell Trench.

western Montana to an elevation of about 4350 feet. When filled to that depth, the lake contained approximately 500 cubic miles of water. Glacial Lake Missoula was, so far as anyone knows, the world's largest ice-dammed lake.

Ice floats. Ice dams float when the water behind them gets deep enough, and then they break up into pieces and the lake drains, catastrophically. When its ice dam floated and broke, Glacial Lake Missoula thundered down Rathdrum Prairie and the valley of the Pend Oreille River, across a broad reach of eastern Washington, finally to the Pacific. Those were the Spokane floods, the greatest of known geologic record.

Imagine a wall of water 2000 feet high, with 500 cubic miles of water behind it. Those horrendous floods surged across eastern Washington, filling each broad valley, then pouring across the drainage divide into the next. The surging water scoured channels through the pale wind blown dust of the Palouse Hills and bit deep into the black basalt beneath to erode the Channeled Scablands of eastern Washington. The floods finally entered the Columbia River through the Wallula Gap near Pasco, Washington.

Thence the Spokane floods descended the Columbia Gorge, filling it to depths as great as 1000 feet, flooded the Willamette Valley of Oregon, and finally emptied into the Pacific Ocean. Everywhere those muddy floods slowed or paused, they left deposits of sediment to record their happening. Everywhere that water went, it carried ice that went aground, then left its litter of ice-rafted boulders where it finally melted. You can trace the passages of the floods by looking for pieces of Belt sedimentary rocks that could have come only from the mountains of northern Idaho and western Montana.

But the ice dam at Pend Oreille Lake was only a tiny southern extremity of the regional ice that covered most of British Columbia, a prong of ice that pushed south down the Purcell Valley. Ice continued to move south after the flood passed, and soon dammed the Clark Fork again. Then another Glacial Lake Missoula formed, deepened until it floated the ice dam, and poured across the Columbia Plateau in another great Spokane flood.

Each of those floods backed up the streams that flow into the Columbia, reversed their flow for a few days, and left its own deposit of sediment in their valley floors. An internal gradation from coarse debris at the base to finer material at the top sharply demarcates each of those flood layers, clearly separating it from those above and below. At least 41 of those flood deposits exist in some of those stream valleys, plainly recording at least that many catastrophic drainages of Glacial Lake Missoula.

Cabinet Gorge. The Pend Oreille River eroded this narrow slot during and after the emptyings of Glacial Lake Missoula.
—U.S. Geological Survey

63

Annual layers in lake sediments deposited in the valleys of western Montana show that the first filling of Glacial Lake Missoula lasted 58 years, and that each successive filling lasted fewer years than the one before. So it seems that the earliest version of the lake was the largest, and that each successive filling accumulated a smaller volume of water than the one that went before. The flood deposits in eastern Washington reveal a similar record: the lowest is the thickest, and they become progressively thinner upward. The pattern is consistent.

To explain that pattern, we suggest a scenario that has the glacier first impounding Glacial Lake Missoula just as the ice age reached its climax, when the ice was at its maximum thickness. Then the climate changed, and the ice age began to wane. As the glacier melted and thinned, it floated in progressively shallower water depths, after fewer years of lake filling. Each successive lake lasted a few years less than the one before, and released a lesser version of the Spokane flood when it finally broke its ice dam.

Most of the known geologic record of Glacial Lake Missoula probably survives from the last ice age, the one that reached its maximum and began to wane about 15,000 years ago. But scattered evidence exists of an earlier and more magnificent version of Glacial Lake Missoula that existed during the ice age of some 100,000 years ago. That part of the story still awaits investigation.

Interstate 90:
Spokane — Lookout Pass
96 miles

Spokane is on the Columbia Plateau; Coeur d'Alene lies at its eastern margin. All the bedrock between is basalt lava flows erupted approximately 15 million years ago. You don't see much of that basalt because it is fairly well buried under glacial debris laid down as the last ice finally melted. That happened long after the final emptying of Glacial Lake Missoula sent the last Spokane flood thundering through. If the basalt were exposed, it would undoubtedly show the same kind of flood scouring you see in the Channeled Scablands of Washington.

All the rocks exposed along the road between the east end of Coeur d'Alene Lake and Lookout Pass are Precambrian sedimentary formations, Belt rocks. Except for some diabase sills and a few small and widely scattered intrusions of granite, the same rocks form all the mountains near the road. Look at these somber rocks closely and with due respect; gray and unassuming as they may be, they contain most of Idaho's mineral wealth.

The Feather Edge of the Columbia Plateau

Several road cuts north of the highway near the east end of Coeur d'Alene Lake expose the thin eastern edge of the Columbia Plateau, where Miocene basalt lava flows lap onto the ancient rocks of the North American continent. Watch for dark brown and black basalt lying on pale gneiss and schist. In some exposures the basalt breaks

Columnarly jointed basalt lava flows lying on weathered metamorphic rocks north of Coeur d'Alene Lake.

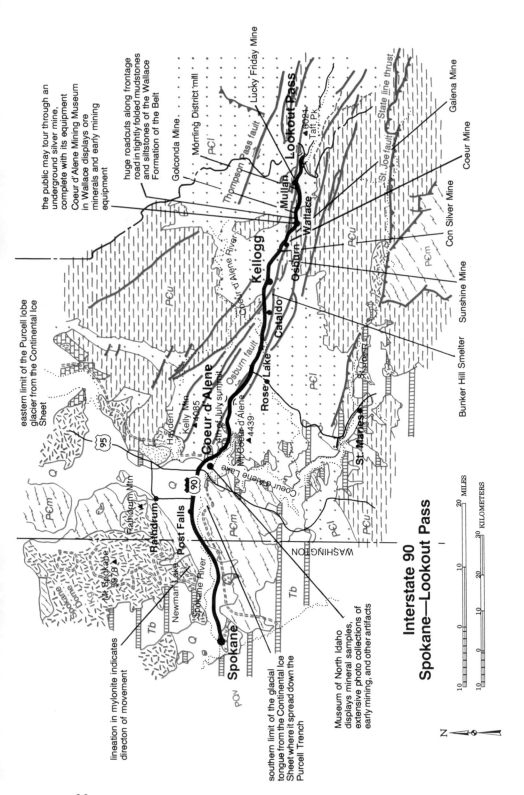

the public may tour through an underground silver mine, complete with its equipment

Coeur d'Alene Mining Museum in Wallace displays ore minerals and early mining equipment

huge roadcuts along frontage road in tightly folded mudstones and siltstones of the Wallace Formation of the Belt

eastern limit of the Purcell lobe glacier from the Continental Ice Sheet

lineation in mylonite indicates direction of movement

southern limit of the glacial tongue from the Continental Ice Sheet where it spread down the Purcell Trench

Museum of North Idaho displays mineral samples, extensive photo collections of early mining, and other artifacts

Golconda Mine

Morning District mill

Lucky Friday Mine

Lookout Pass

Taft Pk.

State line thrust

Galena Mine

Coeur Mine

Con Silver Mine

Sunshine Mine

Bunker Hill Smelter

Thompson Pass fault

Mullan

Wallace

Osburn

St. Joe fault

Kellogg

Coeur d'Alene River

Cataldo

Osburn fault

Rose Lake

St. Joe R.

St. Maries

Coeur d'Alene

Hayden L.

Kelly Mtn

4 July summit

Mt Coeur d'Alene

Coeur d'Alene Lake

Post Falls

Rathdrum

Rathdrum Mtn

Spokane Dome

Mt. Spokane

Newman Lake

Spokane River

Spokane

WASHINGTON

Interstate 90
Spokane—Lookout Pass

N

MILES

KILOMETERS

into vertical columns; in others it is a rubble of irregularly broken cobbles. The nondescript weathered material between basalt and older rock is ancient soil, buried for 15 million years.

Belt Rocks

Most of the Belt rocks along the road between Coeur d'Alene Lake and Lookout Pass are the older formations that appear to have been deposited in fairly deep water. Others are the younger Belt formations laid down in shallow water or on dry land. All are at least a billion years old.

The deep water formations tend to run to dark gray mudstones and muddy sandstones. They generally weather to various shades of rusty brown because they contain scattered crystals of iron pyrite that break down into iron oxides, extremely good pigments. These rocks typically contain little or no evidence of wave motion or exposure to the air, but have the general look of sediments laid down from clouds of muddy water that billow across the deep sea floor.

The younger Belt sedimentary formations are brightly colorful in shades of red, pink, and green. Most are full of little sand ripples that tell of gently oscillating waves, and of the patterns of little polygons that form in suncracked mud. They also contain the fossil remains of algae that must have lived in the bright light of shallow water.

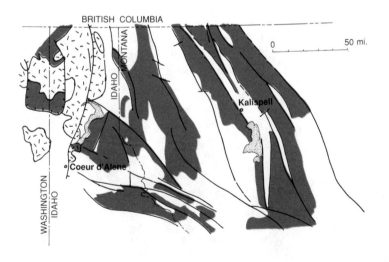

The general distribution of older and younger Belt formations. Their tendency to line up in belts that trend north reflects the deformation that created the northern Rocky Mountains.

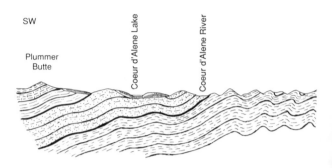

SW

Plummer
Butte

Coeur d'Alene Lake

Coeur d'Alene River

Section across the line of the highway

The Lewis and Clark Fault Zone

The entire route follows the Lewis and Clark fault zone, a broad and enigmatic swarm of faults that trend generally from northwest to southeast. The remarkably straight valley of the Coeur d'Alene River, with the highway beside it, follows the Osburn fault all the way from Coeur d'Alene to Lookout Pass. The crushed rock along large faults is especially vulnerable to weathering and erosion, therefore likely to become a straight river valley.

Giant fault zones are especially hard to analyze. They are so big that they generally reach far beyond the limited area any geologist can easily study, and they tend to manifest themselves differently as they break through different rocks. So it is no surprise to find that

Folded gray mudstones in the Wallace formation, just east of Wallace.

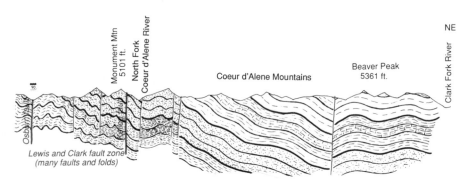

east of Coeur d'Alene Lake.

geologists who work on different parts of giant fault zones often differ in their interpretations.

All geologists who have examined the faults of the Lewis and Clark zone agree that they moved laterally, instead of vertically. Most agree that the zone extends at least as far west as Spokane, at least a few miles east of Missoula; some believe they can trace it west to the Okanogan Valley, east as far as Billings. Very few geologists agree in their ideas of which way the faults moved, and when.

We contend that the Lewis and Clark fault zone has seen two major periods of movement: one about 70 million years ago during late Cretaceous time, the other sometime after about 50 million years ago. The earlier movement bent and offset folds associated with emplacement of the Idaho granite batholith a few miles south of the Lewis and Clark fault zone. Later movements offset at least one small granite intrusion that crystallized about 50 million years ago.

We think the first period of activity saw the faults slipping in a left lateral sense — imagine yourself standing near the fault and looking across it to see the opposite side moving to your left. In other words, the north side of each fault moved west. That is the direction in which some of the faults offset the folds associated with the Idaho batholith. And those must have been the movements that made the old western margin of the continent jog where it crosses the Lewis and Clark fault zone.

We think the second spasm of movement after 50 million years ago moved at least some of the faults in the Lewis and Clark zone in a right lateral direction, back the way they had come. Geologists who do

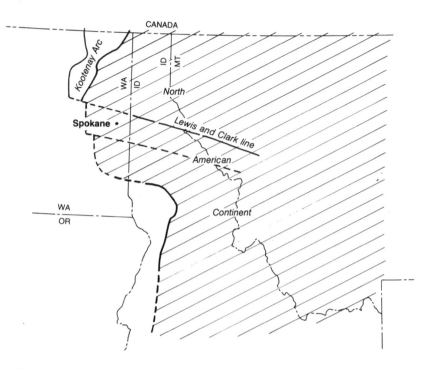

The early period of movement on the Lewis and Clark fault zone offset the old continental margin, giving it the kind of jog you would expect if the rocks north of each fault were moving west.

detailed work in the Coeur d'Alene mining district have insisted for years that they find folds broken and offset in that direction. Much simpler evidence to support that interpretation exists in the Gem stock, a small granite intrusion that intruded the Belt formations north of Wallace about 50 million years ago. A fault sliced off its end and moved the piece in the right lateral direction.

The Coeur d'Alene Mining District

All the way between Smelterville and Mullan, the highway passes through the heart of the Coeur d'Alene district, past one historic mine, mill, smelter, and mining town after another. Nearly every side road leads to still others.

Like most Rocky Mountain mining districts, the Coeur d'Alene began as a placer gold discovery, in 1881. The customary nineteenth century horde of otherwise unemployed gold miners promptly stampeded into the area, and a town named Eagle City materialized as though by magic. The gold mining never did amount to much. Neither, for that matter, did Eagle City, which collapsed almost as quickly as it rose.

Lead and silver deposits were first discovered in 1884, most of those in the district by 1886. Local folklore has it that a jackass named Old Bill was so transfixed by the gleam of metal in an outcrop that he led Noah Kellogg to discover the Bunker Hill and Sullivan ore body in 1885. The more prosaic fact is that part of the fateful outcrop survives to assure us that it is an exposure of rusty sandstone utterly without bright metallic gleam. Maybe Old Bill liked the look of rusty sandstone.

Regardless of how the ore was discovered, the Coeur d'Alene district went on to become the biggest in Idaho, one of the several biggest in North America. Production from the Coeur d'Alene mines accounts for more than 80 percent of all the ore ever mined in Idaho, far more than the combined output of all the other mining districts in the state. The actual figures are numbers of the sort that strain the imagination, like those that describe the national debt, geologic time, and astronomic distances.

In the century between 1884 and 1984, the Coeur d'Alene mines produced 507,300 ounces of gold — not bad for a district that recovers gold mainly as a by-product. Meanwhile, the district produced just under a billion ounces of silver — the billionth came up the shaft in 1985. All that silver was accompanied by some 31,000 tons of lead, 4,000 tons of copper, and 3,300 tons of zinc. It is virtually impossible to assign a meaningful dollar value to production from the Coeur d'Alene district because the prices of the various metals have fluctuated so greatly during the last century, as has the value of the dollar.

The district produces about 40 percent of all the silver mined in the country. Most of it comes from a mineral called tetrahedrite, which also contains quite a bit of copper and sulfur, along with some antimony. Tetrahedrite is a rather nondescript dark gray mineral with hardly a trace of any metallic gleam, the sort of thing only a miner could love, definitely not a mineral that would catch Old Bill's discriminating eye.

But a bit of galena might have caught his attention. Lead, along with another bit of silver, comes mostly from galena, a lead sulfide mineral that typically crystallizes into dark gray cubes with a brightly metallic luster. Lead atoms have enough in common with those of silver that the two can substitute for each other in galena. If the mineral contains a lot of silver, the cubic crystals acquire slightly curving faces. That happens because silver atoms are not exactly the same size as those of lead, so their substitution in the crystal distorts it, just as substitution of occasional smaller bricks for the standard size would distort the shape of a wall.

All those valuable ores exist in Precambrian sedimentary rocks, the Belt formations that appear in all the road cuts between Coeur d'Alene and Lookout Pass. They occur in steeply dipping veins associated with the faults of the Lewis and Clark fault zone. Some of the mines have pursued those veins to great depth.

The Star Mine near Burke, for example, reached a depth of some 8300 feet before it closed in 1982, and was in its time the deepest lead and zinc mine in the world. The Lucky Friday Mine at Mullan has the deepest single shaft in the world outside of South Africa; it sinks 6200 feet in a continuous drop. The Sunshine Mine west of Wallace opened in 1884, and was producing about 4,500,000 ounces of silver per year a century later from depths as great as 7000 feet.

Despite its more than a century of rich production, the Coeur d'Alene district is far from exhausted. Someday, of course, all the ore will be gone, but so far, geologists have been able to find new ore bodies as the old ones were worked out. During the late 1970s and early 1980s, the mines produced an average of 14.5 million ounces of silver every year, along with proportionate amounts of the other metals. That kind of production should continue many years into the future, provided that metals prices remain at least as high as they were during the early 1980s.

Coeur d'Alene Lake

Coeur d'Alene Lake is an oddity: a glacial lake that floods country that was never glaciated. During the great ice age of some 70,000 to 130,000 years ago, a glacier poured south from Canada the full length of the Purcell Valley. At its maximum extent, that ice passed all the way through Rathdrum Prairie almost to the line of the highway, where it left a terminal moraine of glacial till and deep deposits of outwash. Those deposits dammed the St. Joe River, which formerly continued north through Rathdrum Prairie. The result is Coeur d'Alene Lake. It floods a part of the St. Joe River Valley just beyond the farthest reach of the ice.

U.S. 2:
Newport — Sandpoint
24 miles

Idaho 200:
Sandpoint — Montana Line
32 miles

Between Newport and Sandpoint, the road follows the Pend Oreille River below Pend Oreille Lake. The fancy rocks exposed in the long series of magnificent road cuts between Priest River and Laclede are mostly schist and granitic gneiss, the Priest River metamorphic complex. Most of the pale gray bedrock exposed between Laclede and Sandpoint is massive granite, the Kaniksu batholith.

East of Sandpoint, the road follows the north side of Pend Oreille Lake, and the Clark Fork River to the state line. Virtually all the bedrock exposed along this part of the route is Precambrian sedimentary formations, Belt rocks. In a few places, those formations appear as spectacular road cuts, but in most areas a deep cover of glacial sediment buries them.

Belt Rocks

Except for occasional glimpses of black diabase sills, all the outcrops you see east of Sandpoint expose Precambrian sedimentary rocks, Belt formations. The several formations in this area range from dark gray, almost black, muddy sandstones through various greenish shales and mudstones to red mudstone and sandstone.

The dark gray mudstones lack fossil algae and mudcracks, show no sign of having been exposed to the air or to sunlight as they were deposited. Neither do they contain ripple marks to show that they felt much wave action. The absence of those influences suggests that these somber formations may well have accumulated in water deep enough to put the bottom out of reach of light or waves. They look like turbidites, the kind of muddy sandstones laid down beneath clouds of muddy water that sweep down long slopes, then billow out across the deep sea floor.

The colorful red and green mudstones and red sandstone were deposited in shallow water and on land. Many of the bedding surfaces in those formations show the characteristic polygonal patterns of sun-cracked mud. Others are covered with intricate patterns of little ripple marks, the signatures of small waves oscillating in shallow

73

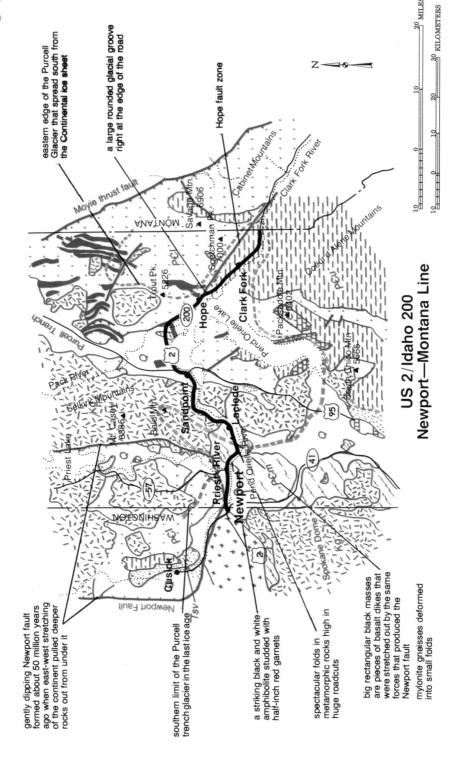

US 2/Idaho 200
Newport—Montana Line

eastern edge of the Purcell Glacier that spread south from **the Continental ice sheet**

a large rounded glacial groove right at the edge of the road

Hope fault zone

Movie thrust fault

Cabinet Mountains

Clark Fork River

Clark Fork

Savage Mtn. 5906

Scotchman Pk. 7000

MONTANA

PCU

Trout Pk. 5226

200

Hope

Clark Fork

Packsaddle Mtn. 6402

Coeur d'Alene Mountains

PCU

Purcell Trench

Pack River

Selkirk Mountains

Mt. Casey 6385

Bald Mtn.

Sandpoint

Pend Oreille Lake

South Chilco Mtn. 5665

Laclede

95

2

Priest Lake

57

Priest River

Pend Oreille River

Newport

PCU

WASHINGTON

41

Spokane Dome

Kg

Cusick

Newport Fault

Tsv

gently dipping Newport fault formed about 50 million years ago when east-west stretching of the continent pulled deeper rocks out from under it

southern limit of the Purcell trench glacier in the last ice age

a striking black and white amphibolite studded with half-inch red garnets

spectacular folds in metamorphic rocks high in huge roadcuts

big rectangular black masses are pieces of basalt dikes that were stretched out by the same forces that produced the Newport fault

mylonite gneisses deformed into small folds

N

MILES
KILOMETERS
20
30
10
10

pools of water. Occasional exposures even contain stromatolites, peculiar structures made by extremely primitive blue-green algae. Large stromatolites usually remind us of cabbages, smaller ones of clustered brussels sprouts preserved in the rock. In fact, nothing could be more far fetched; identical stromatolites still grow here and there today, and there is no doubt that they are algae of the most primitive kind. Those extremely primitive algae were green plants that made their tissues through photosynthesis. Their fossil remains in these ancient rocks tell of bright light streaming through shallow water.

The Belt rocks are full of black diabase sills, along with a few dikes. They are hard to spot because their color is not distinctive enough to make them conspicuous in a region so full of dark gray Belt formations, and they do not form especially prominent outcrops. A few reveal themselves through a weakly developed tendency to break into columns similar to those that mark basalt lava flows.

Kaniksu Batholith

The great mass of granite that forms the Selkirk Mountains northwest of Sandpoint and also extends well south of the highway is the Kaniksu batholith. Watch for the exposures of massive granite along the road between Laclede and Sandpoint. The Kaniksu batholith formed approximately 70 million years ago, at the same time the Idaho batholith was forming in central Idaho, and for the same reason.

The Priest River Metamorphic Complex

Connoisseurs of fine road cuts will appreciate the magnificent series that stretches along the Pend Oreille River for several miles between Priest River and Laclede. They expose the granites under the

The pale gneiss in this road cut formed through recrystallization of pre-Belt granite; the black oblongs appear to be rotated remnants of a dismembered diabase dike. The vertical lines are drill holes

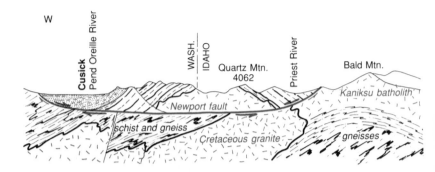

Section across the Idaho panhandle about at Sandpoint showing those formed at shallower depth to the east.

Priest River metamorphic complex, another of those formations that will be much easier to understand after the rocks of this part of Idaho have received more attention.

One interpretation views the metamorphosed sediments of the Priest River complex as an expanse of Precambrian basement rock, ancient continental crust, that was exposed when the younger rocks that once covered it moved east. Another regards the Priest River complex as Belt formations intensely metamorphosed while still more intense heating farther north melted the magma that eventually became the Kaniksu batholith. We prefer the second interpretation, but know of no evidence that would eliminate the first. In the present state of knowledge, both views are reasonable.

Whatever the origin of the rocks, the high road cuts through the old granite and the Priest River complex provide one of the best views of rocks thoroughly metamorphosed at high temperature that you are likely to see, anywhere. It is extremely fortunate that these cuts in the Priest River metamorphic complex were made years ago, before it became customary to desecrate new road cuts by reducing them to a gentle angle, covering them with dirt, and planting grass. Had that been done here, Idaho would have lost one of its finest geologic spectacles.

The Purcell Trench

The highway crosses the Purcell trench as it passes along the north side of Pend Oreille Lake. As you look either way across the Purcell trench from the area near Sandpoint, try to imagine the rocks in the

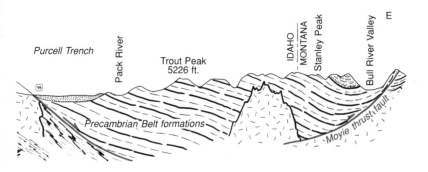

how the Purcell trench separates deep seated rocks to the west from

mountains to the east moving off those in the mountains to the west, opening the trench behind them. Quite a strain, isn't it? Nevertheless, it is the easiest way to understand the rocks in those mountains. Everything fits.

First, consider the granite of the Kaniksu batholith in the Selkirk Mountains along the west side of the Purcell trench. It certainly formed at least several miles below the surface. And remember that the eastern face of that granite is mylonite, a strongly sheared rock that forms where faults penetrate very hot rocks deep beneath the surface. Then remember that the rocks in the Cabinet and Purcell Mountains on the east side of the Purcell trench are Belt formations, rocks that properly belong at much shallower depths than the granite a few miles to the west. It does seem logical to imagine exposing that granite by moving the Belt formations off it, shearing the mylonite in the process.

The Hope Fault

The Hope fault trends parallel to the road east of Sandpoint. It starts at, but does not offset, the Purcell trench, which means that it is either the same age as the Purcell trench, or older. The Hope fault extends southeast more than a hundred miles into western Montana, where it finally angles into the Lewis and Clark fault zone, which crosses Idaho along the line of Interstate 90.

Although the picture is not completely clear, the weight of the evidence suggests that the rocks north of the Hope fault moved southeast relative to those on its south side. Perhaps the Hope fault

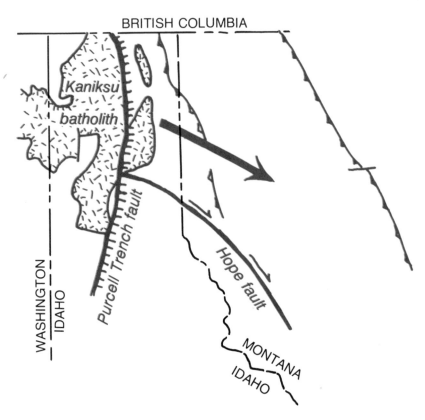

One possible interpretation of the Hope fault as the southern boundary of the slab that moved off the Kaniksu batholith.

is the torn southern edge of the slab that moved eastward off the Kaniksu batholith west of the Purcell trench.

It would help if the timing of the Hope fault were not so hopelessly obscure. All the rocks on both sides of the fault are Belt formations a billion or more years old. Nowhere along its considerable length does the fault offset younger rocks, and the youngest rocks that cover it without offset are glacial deposits. Evidence that merely brackets the age of the fault somewhere between one billion and ten thousand years ago is no help.

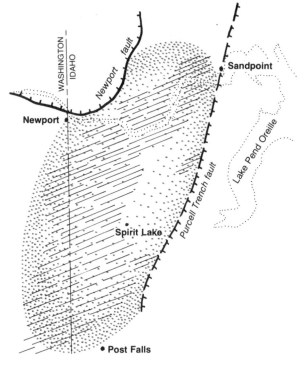

The Spokane dome with its mylonite.

Spokane Dome

The expanse of granite that fills most of the geologic map between Newport and Spokane is the Spokane dome, another enigmatic mass of rock. There is little doubt that the granite is about 70 million years old, the same age as the Kaniksu batholith and many other large masses of Idaho granite. A prominent zone of strongly sheared rock, mylonite, lies along its eastern margin, which is the western edge of the Purcell trench. The mylonite extends over the top of the dome. And down its western flank.

A close view of the sheared mylonite on the east side of the Spokane dome.

79

Detailed studies of the arrangement of mineral grains in the mylonite seem to show that the great thickness of rock that covered it was moving east as it sheared the mylonite. It seems reasonable to suggest that the eastward moving cover was the rocks that now form the mountains east of the Purcell trench south of Sandpoint. If so, then the covering slab must have extended west of the Spokane dome, or else the mylonite would not extend down its western flank. The deep rocks of the Spokane dome rose as they lost their heavy cover of sedimentary rocks.

The Newport Fault

We don't know of any other fault quite like the Newport fault. It looks on the geologic map like a big loop open to the north, which almost closes on itself to make a circle, but definitely does not. Rocks within that open loop are younger than those outside, and formed at shallower depth. They must have dropped. The textures of the weakly sheared mylonites along the fault indicate that the rocks beneath it pulled out from under the younger ones within, letting them drop. Evidently, the earth's crust was stretching when the fault formed, about 50 million years ago.

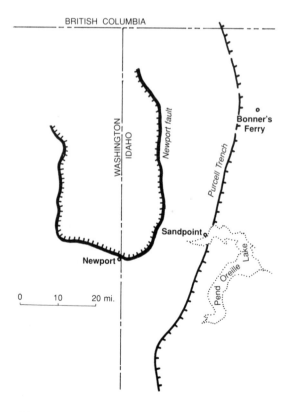

The Newport fault suggests a horseshoe laid on the geologic map, half in Washington, half in Idaho, the open end facing north.

W   E

Newport fault

Section through the Newport fault showing how the rocks within it dropped as those on either side pulled apart. The dashed lines show schematically how this movement may fit into a broader view of how the earth's crust stretches.

In Washington, the loop of the Newport fault encloses a patch of sedimentary rocks called the Tiger formation, which contains fossils of animals that lived during Eocene time, about 50 million years ago. It seems reasonable to infer that the rocks within the Newport fault were dropping then to open the shallow basin that holds the Tiger formation.

Many geologists now suspect that the earth's crust stretches by breaking into segments shaped like big lenses that slide past each other like so many pumpkin seeds in a bag. If so, the Newport fault is an excellent illustration of how that might work. The Spokane dome may illustrate the other side of the process, a place where deep rocks appeared at the surface as the crust stretched.

Glaciation

When the ice ages were at their maximum, most recently about 15,000 years ago, ice filled the valleys of northern Idaho, leaving only the higher hills standing above it. The highest mountains caught enough snow to breed glaciers of their own that poured down the valleys to join the groaning sea of ponderously moving ice below.

Although the glaciers of the most recent ice age were not as big as some of their predecessors, their effects are the best preserved, the ones you are most likely to see. The most obvious souvenirs of their passage are erratic boulders, blocks of rock the ice trucked in from somewhere north, then left as it melted at the end of the ice age. In this region, almost any isolated boulder not attached to bedrock is likely to be a glacial erratic.

Glacially transported boulder in the woods near Sandpoint.

Glacial ice, especially that near the base of the glacier, is full of mud, sand, pebbles, and boulders. All that hard debris scrapes, polishes, and scratches any exposed bedrock the glacier may cross. Now that the ice is gone, we see slick bedrock surfaces covered with parallel scratches that precisely record the direction of ice movement — if the scratches make a criss-crossing pattern, you are probably looking at a glacially transported boulder that shifted position in the ice. Detailed maps that show the directions of glacial scratches in bedrock vividly portray how the ice moved down valleys, bulged into side valleys, and in places wrapped around the ends of ridges.

Glacial Lake Missoula flooded the Clark Fork Valley east of Pend Oreille Lake, at times to depths of 2000 feet. Powerful currents that flowed as the lake drained swept most of the line of Highway 200 free of lake sediment. But patches of those deposits did survive here and there. Watch for road cut exposures of soft silt that is slightly tan in dry weather, very pale pink on rainy days.

Rocks embedded in passing glacial ice carved these giant grooves in this exposure of Precambrian sedimentary rock beside the road near Clark Fork.

82

U.S. 95:
Coeur d'Alene — British Columbia
111 miles

The road between Coeur d'Alene and Careywood crosses the broad floor of Rathdrum Prairie, the southern end of the Purcell trench. In the area a few miles on either side of Careywood, the road edges into the Selkirk Mountains on the west side of the Purcell trench, where it crosses granite, a southern outpost of the Kaniksu batholith.

North of Sandpoint, the road follows the eastern side of the Purcell trench with the Selkirk Mountains rising on the western horizon. The Cabinet Mountains lie east of the road along the southern part of that drive. The Purcell Mountains rise east of the road north of Bonners Ferry. Between Copeland and the Canadian Border, the road winds through the Purcell Mountains.

The Purcell Trench

The Purcell trench is a straight topographic trough that angles far north into British Columbia, where it eventually merges with the Rocky Mountain trench, which continues far north into the Yukon. From Sandpoint north, the Purcell trench is sharply obvious, almost impossible to overlook or ignore. South of Sandpoint, it seems to unravel into a diffuse pattern of valleys, of which Rathdrum Prairie is the largest and most sharply defined. The Purcell trench is much too straight, much too wide, and much too long to have formed as an erosional valley. It is clearly a major bedrock structure strongly expressed in the landscape.

The Cabinet and Purcell mountains east of the Purcell trench consist mostly of Belt sedimentary formations. They appear to have moved east off the granite and metamorphic rocks of the Selkirk Mountains to the west, leaving the Purcell trench as an open gap behind the trailing edge of the moving slab. That movement uncovered the deep-seated rocks of the Selkirk Range as you would if you pulled the blankets off a bed to reveal the lower sheet. A strongly developed shear zone, a mylonite, poorly exposed in places above the glacial debris that covers most of the eastern flank of the Selkirk Mountains appears to be the fault that carried the movement.

During the wet period of late Miocene time, a large river flowed south through the Purcell trench into an even larger stream near Spokane. Then the enormous lava flows that built the Columbia

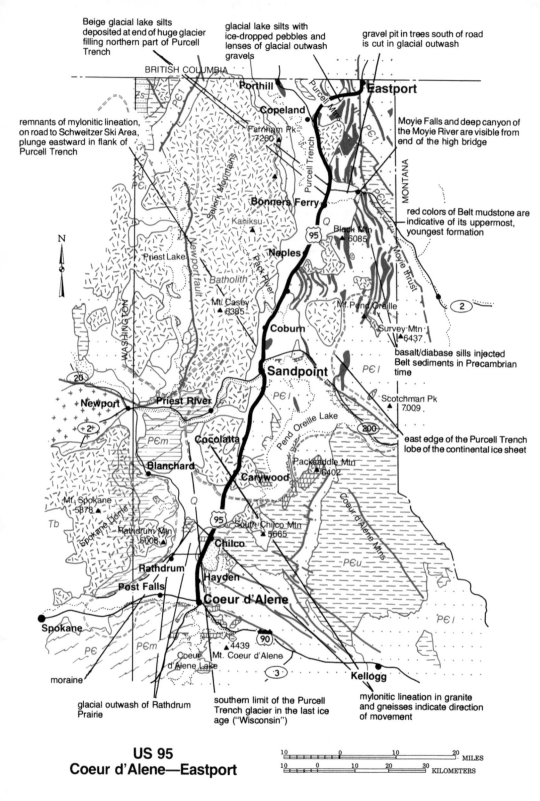

Beige glacial lake silts deposited at end of huge glacier filling northern part of Purcell Trench

glacial lake silts with ice-dropped pebbles and lenses of glacial outwash gravels

gravel pit in trees south of road is cut in glacial outwash

BRITISH COLUMBIA

Porthill

Eastport

Copeland

remnants of mylonitic lineation, on road to Schweitzer Ski Area, plunge eastward in flank of Purcell Trench

Farnham Pk 7260

Moyie Falls and deep canyon of the Moyie River are visible from end of the high bridge

MONTANA

Bonners Ferry

red colors of Belt mudstone are indicative of its uppermost, youngest formation

Kaniksu

95

Black Mtn 6085

Priest Lake

Naples

2

Mt Casey 6385

Mt. Pend Oreille

Coburn

Survey Mtn 6437

basalt/diabase sills injected Belt sediments in Precambrian time

Sandpoint

PЄl

WASHINGTON

Scotchman Pk 7.009

20

Newport

Priest River

PЄm

200

east edge of the Purcell Trench lobe of the continental ice sheet

2

Cocolalla

Blanchard

Packsaddle Mtn 6402

Carywood

Mt. Spokane 5878

Tb

95

South Chilco Mtn 5665

Rathdrum Mtn 5008

PЄu

Chilco

Rathdrum

Hayden

Post Falls

Coeur d'Alene

PЄl

Spokane

PЄ

PЄm

90

4439

Mt. Coeur d'Alene

Coeur d'Alene Lake

3

Kellogg

moraine

glacial outwash of Rathdrum Prairie

southern limit of the Purcell Trench glacier in the last ice age ("Wisconsin")

mylonitic lineation in granite and gneisses indicate direction of movement

US 95
Coeur d'Alene—Eastport

10   0   10   20 MILES
10   0   10   20   30 KILOMETERS

84

Plateau about 15 million years ago backed into Rathdrum Prairie to impound Miocene Lake Rathdrum, which flooded much of the Idaho panhandle portion of the Purcell trench. Layers of sediment record the lake. Ice age glaciers left the entire Purcell trench so deeply mantled with glacial debris that exposures of the lava flows and the lake bed sediments are hard to find.

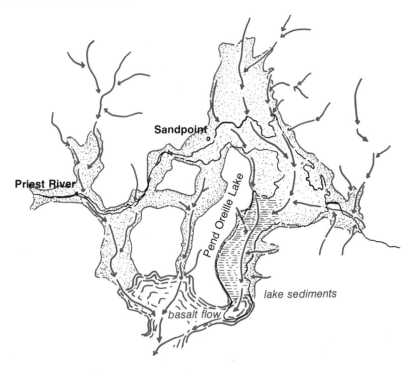

Miocene Lake Rathdrum; the arrows show the stream drainage that flowed through the Purcell trench during that wet time.

The Cabinet and Purcell Mountains

The Cabinet and Purcell mountains east of the highway between Sandpoint and Bonners Ferry consist mostly of Precambrian sedimentary formations full of diabase sills. Both also contain several small intrusions of much younger granite.

We think the masses of granite were originally high protrusions of the Kaniksu batholith into the Belt sedimentary formations above its roof. When that cover of sedimentary rocks moved east to where we now see them in the Cabinet and Purcell mountains, it sliced the small

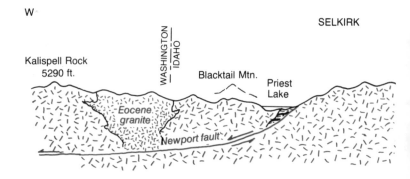

W

SELKIRK

Section across Idaho just south of Bonners Ferry.

masses of granite off the top of the Kaniksu batholith. Although all granites consist of the same minerals combined in about the same proportions, individual granites do have their own distinctive character. You can learn to recognize them on sight. Granites in the Purcell and Cabinet mountains look just like the granite of the Kaniksu batholith in the Selkirk Range west of the Purcell trench.

Lakeview Mining District

Early in the fall of 1888, three prospectors found outcrops of a rich silver vein near the southeastern shore of Pend Oreille Lake. The town of Chloride sprang up almost immediately, reached a population of about 2000 by mid-winter, then wilted as the snow melted. Most of the people were gone long before the next daisies bloomed.

Despite the big exodus of 1889, several small mines worked off and on around Chloride for the next 60 years. Most authorities estimate total production from the district at about two million dollars worth

Section across the highway north of Coeur d'Alene. Rocks east of west. The mylonite west of Rathdrum Prairie is the fault.

W

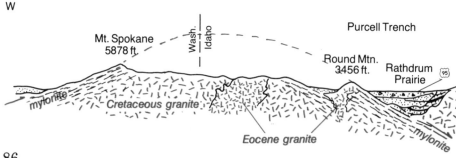

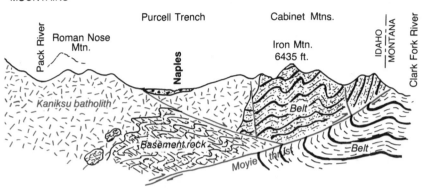

of silver, lead, and zinc — pathetically little to show for all the investment in equipment and all those man years of labor.

The ore, such as it may be, is in veins that fill fractures in Precambrian sedimentary rocks, Belt formations. The metallic minerals probably came from nearby granite intrusions related to the Kaniksu batholith. Hot water solutions circulating through and around the masses of molten magma brought dissolved metals into these cooler rocks, where they precipitated in fractures to form veins.

Too Little Copper in the Purcell Mountains

Precambrian sedimentary rocks in the Purcell Mountains contain dozens of diabase sills, along with a few dikes. Everywhere those igneous intrusions exist, prospectors have found widely scattered small deposits of copper, lead, and zinc in the Belt sedimentary rocks near their contacts. The metallic minerals must have come from the diabase magma. All the many attempts to develop those deposits into

the Purcell trench moved off the Cretaceous granite to the

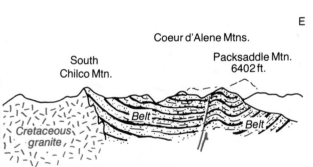

mines have failed, invariably because the deposits are neither large nor rich. The Purcell Mountains contain plenty of copper prospects that inspired great optimism in their time, but produced no ore worth mentioning.

Idaho's Craziest Mine

Some appropriately bizarre monument should be placed to mark the scene of one of the most grotesque schemes in the entire eccentric history of early western mining. A prospector in the last stages of mercury dementia might be the proper subject. The proper place is at the mouth of Boulder Creek, which is southeast of Bonners Ferry almost to the Montana line, about three miles south of Highway 2. The comedy ran its course during the early years of this century.

In the first stage of the scheme, a mining company built a sort of giant sieve in the bottom of the canyon. It was made of rails lined up side by side with gaps between them, like an enormous cattle guard that covered the entire canyon floor. The rails ran parallel to the valley walls for a distance of some 800 feet.

When the oversized sieve was ready, the miners used water brought in on long flumes to wash 800,000 cubic yards of gravel from high banks onto the canyon floor upstream. That is a lot of gravel. Then they opened a dam still farther upstream to wash the 800,000 cubic yards of gravel across the sieve. The coarse gravel was supposed to roll over the sieve and down the stream while the finer material that might contain gold settled through the sieve into a basin beneath. After the big flush, the company planned to pump the fine material out of the basin and concentrate the gold. Nothing worked.

Rolling boulders wrecked the sieve almost immediately, so the scheme never actually went into full operation. But that didn't really matter; the project would have failed miserably even if the giant sieve had worked. So far as anyone knows, the gravel is certifiably gold free. Evidently everyone involved just cheerfully assumed that the gold was there, and no one bothered to check. You can still see big piles of that barren gravel in the bed of Boulder Creek.

Glaciation

During both of the ice ages that left a clear mark on the landscape, ice filled the valleys of British Columbia and buried the lower mountains to make a regional ice sheet. That general ice cover reached far enough south to bury most of the northern part of the Idaho panhandle, and sent long fingers of ice farther south along the

Pebbles and cobbles scattered through layers of sand in a deposit of glacial outwash exposed beside the highway a few miles south of the Canadian border.

major valleys. Meanwhile, valley glaciers carved the higher mountains that stood above the level of the regional ice cover, or beyond its southern reach.

During the earlier and larger ice age of about 100,000 years ago, ice moved south through the Purcell trench all the way to Coeur d'Alene and Spokane. Smaller glaciers of the last ice age barely reached south of Pend Oreille Lake. Between them, the glaciers left the floor of the Purcell trench so deeply mantled with glacial debris that good bedrock exposures are few.

Pend Oreille Lake is more than 1000 feet deep in some places. Evidently, the thick glacier of the earlier ice age scoured the deep accumulations of soft sediment deposited in Miocene Lake Rathdrum. It is hard to imagine that the glacier of the last ice age could have eroded so deeply in an area so close to its southernmost end, where it was already beginning to thin.

As the last ice age ended, a large mass of dead ice must have lingered in the deep basin of Pend Oreille Lake long after glaciers elsewhere in the neighborhood had melted. Otherwise, meltwater would surely have swept in enough sand and gravel to fill the hole level with the rest of the valley floor. When that last ice finally melted, it bequeathed the deep basin clear of sediment to Pend Oreille Lake.

The Cabinet Mountains were south of the regional ice cover, but high enough to snatch plenty of snow out of the ice age clouds. They spawned large valley glaciers, which left them a sea of craggy peaks rising above deeply gouged valleys. The high parts of the range contain magnificent alpine landscapes, nearly every scene eminently suitable for a calendar. Meanwhile, regional ice moving south from British Columbia nearly buried the Purcell Mountains, rasping the entire range as though it were a sandpapered block of wood. Now that the ice is gone, the Purcell Mountains have a smoothly rounded look quite unlike the ragged skyline of the Cabinet Mountains.

Glacial Lake Missoula

The long drive through Rathdrum Prairie follows the main drainage route for the torrent of water released when the ice dam at the present site of Pend Oreille Lake floated and burst, suddenly draining Glacial Lake Missoula. Those catastrophes came one after the other as the advancing glacier repeatedly impounded a new lake and then washed away as the water rose high enough to float the ice. The passage of those floods scoured the Channeled Scablands across a wide swath of eastern Washington, yet the floor of Rathdrum Prairie shows no evidence of flood scouring. Why?

Many people who object to the argument that Glacial Lake Missoula drained catastrophically point to the abscence of much evidence of torrential flooding in Rathdrum Prairie. How, they ask, could such great floods have passed that way without eroding the floor of the prairie? It is a good question.

The obvious rejoinder lies in the deep deposits of glacial debris laid down in Rathdrum Prairie as the ice finally melted after Glacial Lake Missoula emptied for the last time. Floods of muddy meltwater draining from the rapidly melting ice to the north covered the floor of Rathdrum Prairie with layer after layer of sand, silt, and gravel. Those deposits buried the evidence of flood scouring.

Moyie Falls on the Moyie River connects a pair of pools in Belt mudstones north of U.S. 2, east of Bonners Ferry.

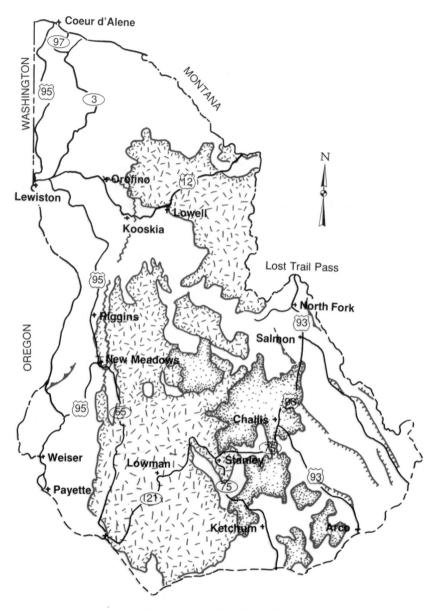

Roads covered in this section.

III
Central Idaho:
The Granitic Core

THE PLATE TECTONIC SCENE

Starting sometime during Jurassic time, perhaps about 150 million years ago, an oceanic trench formed off the old western margin of North America, close to the present western border of Idaho. That happened because North America had just then begun to move west, away from the mid-Atlantic oceanic ridge, which had just just begun to open. As the continent moved west, the floor of the Pacific Ocean slid beneath it, beneath Idaho. Nothing has been the same since. That collision started the long chain of events that created most of the northern Rocky Mountains.

The Early Mountain Welt

It was a ponderous collision, that long and grinding encounter between the lithospheric plates that carry the floor of the Pacific Ocean and the North American continent. One of its first victims was the western edge of the continent, which was thoroughly mashed, compressed as though it were a collapsing accordion.

Slabs of continental rocks jammed together along big thrust faults that cut deep through the crust. Their moving thickened the continent, raising a broad welt along the western margin of North America. We don't know how high that welt was, but it must have been mountainous, probably higher than the mountains you see today. It was the first stage in the development of the northern Rocky Mountains.

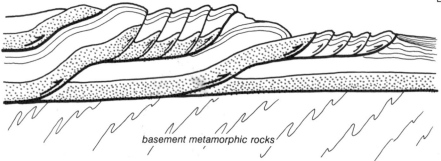

W   E

basement metamorphic rocks

Thrust faults thicken the old western margin of North America.

The Mountains Spread

Certainly, the sinking ocean floor heated as it slid deep beneath the continent; and the rocks in the oceanic crust do lose water as they reach a red heat. So, imagine enormous amounts of steam rising from the sinking ocean floor and reducing the melting temperature of the already hot rocks above it. For tens of millions of years, that steam soaked the mantle above the sinking slab, causing the slab to partly melt to form basalt magma. The rising basalt magma heated the continental crust above that; it was a glowing blow torch playing on the rocks from below. As the old continental crust and the thick accumulations of Belt sedimentary formations that lay on it reached a red heat, they recrystallized to form gneisses and schists.

By late Cretaceous time, large volumes of those ancient crustal rocks were so hot that they melted to form granite magmas, which rose through the crust. That magma eventually crystallized into enormous masses of granite, the rock you now see most of in central Idaho.

Large masses of molten magma rose into the mountainous welt along the edge of the continent, weakening it in the same way that too much soft icing between the layers weakens a cake. Finally, miles of rocks above the granite began to shear off along fractures, which the molten magma promptly filled. Great slabs of rock thousands of feet thick moved east along those fractures, those big thrust faults. They rode on the extremely weak layer of molten rock that lubricated their passage as though it were indeed the soft icing between the

94

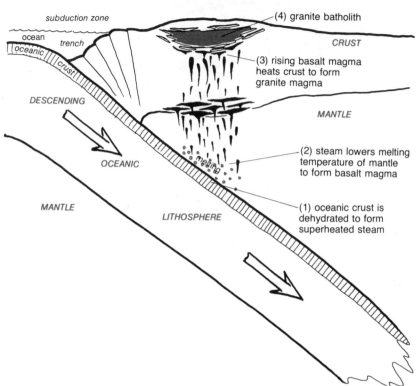

Steam rising from the sinking oceanic crust melts rocks in the overlying mantle to form basalt magma. That basalt heats the lower part of the continental crust to form granitic magma.

layers of a cake. Those slabs, slowly gliding on their sole of magma, spread the early mountainous welt east into a broad zone of eastern Idaho and western Montana, transforming it into the northern Rocky Mountains.

Magma also moved east along the sliding thrust faults. Part of it erupted in southwestern Montana, where it built a long row of volcanoes along the eastern edge of the spreading mountain mass. So far as we know, none of the magma from late Cretaceous time erupted to the surface in Idaho.

The magma that did not erupt finally crystallized into large masses of granite. The largest of those is the Idaho batholith, which fills so much of central Idaho. Smaller masses of granite lie in great sheets along some of the big thrust faults in eastern Idaho and western Montana.

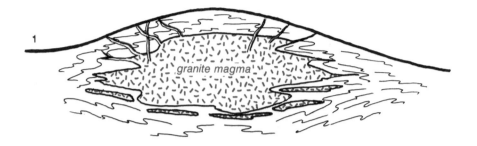

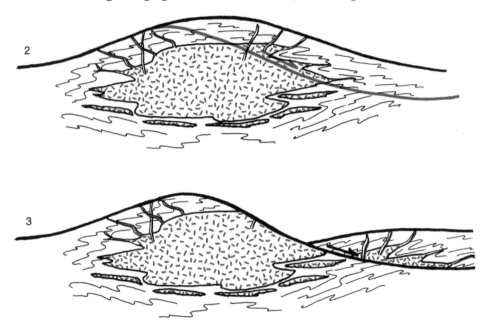

Granitic magma engorges the mountainous welt, weakening the rocks within.

The welt collapses eastward into western Montana.

Eastward movement of the higher parts of the early mountainous welt peeled the cover off the granites of central Idaho, exposing them as they crystallized. After the top moved off the early mountainous welt, the stripped lower part rose simply because the continental crust floats on the mantle; it rose for the same reason that a canoe rises as you step out of it. So the mountains of central Idaho immediately regained approximately three fourths of the elevation they lost when the upper part of the early mountainous welt moved into western Montana.

Meanwhile, the trench off the west coast continued to gobble up the ocean floor at a rate of a couple of inches or so every year. That patient trench eventually reeled islands and groups of islands in out of the Pacific Ocean, gathered them together, and added them one after the other to the western margin of the continent.

The Migratory Islands Come Home

Every inch the mid-Atlantic ridge added to the growing Atlantic Ocean moved North America another inch west, and made the Pacific Ocean an inch narrower. Those inches of ocean floor went down the trench just west of Idaho. Every inch of ocean floor the trench consumed brought the islands in the Pacific Ocean another inch closer to the west coast. That happened at a rate of about two inches every year, more or less.

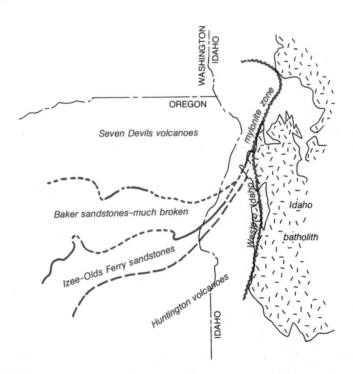

The old western margin of the continent, the mylonite that marks it, and the pieces of continent that docked onto it. —From U.S. Geological Survey Professional Paper 1435

97

Two inches every year translates into a mile every 32,000 years, 31 miles every million years. By about 100 million years ago, the plate collision along the old western margin of North America had been in progress more than 50 million years, time enough to slide more than 1000 miles of oceanic crust under the western margin of North America.

Islands then scattered across the eastern Pacific Ocean included some odd scraps of continental crust, as well as one or more volcanic chains fringed with tropical coral reefs. Most of the rocks in the volcanic islands formed during Permian and Triassic time, something like 200 million years ago. No one knows where they formed. Now that those islands are firmly mashed into the western edge of Idaho, it is no longer clear whether they were one group, or several.

Generation after generation of dinosaurs watched those vagrant islands slowly rise out of the southwestern horizon, come a bit closer every year, and eventually arrive on the old west coast sometime around 100 million years ago, give or take 10 million years. The wandering islands then became much of the western fringe of central Idaho, as well as the Blue and Wallowa mountains of northeastern Oregon. Idaho geologists call the rocks of those exotic islands the Seven Devils and Riggins complexes. It is impossible to know whether the old islands exposed along the western edge of central Idaho continue into the Blue and Wallowa Mountains of Oregon because the intervening plateau basalt buries the connecting rocks. Probably they do.

Basalt lava flows also conceal any connection that may exist between the old islands now embedded in western Idaho and the Okanogan microcontinent of north-central Washington and southern British Columbia. That much larger stray island is a scrap of continental crust that also joined North America sometime around 100 million years ago.

The Old Continental Margin

A sharply curving zone of very strongly sheared rock, mylonite, seems to trace the old continental margin through a stretch of western Idaho at least 80 miles long. That sheared zone generally separates pale gray granites of the Idaho batholith on the east from much darker gray rocks called diorites on the west. The granites are certainly continental rocks, whereas the diorites have strong oceanic affinities.

That distinction between oceanic and continental rocks shows up in several ways. Most obviously, the composition of diorite is much closer to that of the dark rocks of the oceanic crust than to that of the pale granites of the continents. More subtle features lead to a similar comparison. For example, the isotopes of strontium in the granite are like those in continental rocks, whereas strontium in the diorites is like that of oceanic regions. It does seem that the diorite magmas formed as parts of the old oceanic islands melted.

Most of the dark diorites are between about 80 and 100 million years old. The granites east of the mylonite zone crystallized much later, between about 60 and 70 million years ago. The diorites weld the seam between the old continental margin and the former islands to the west, so their age probably records the time when the wandering islands firmly docked onto the continent. The northeastern edge of the diorite is strongly sheared next to the old continent. That shearing stopped as the diorite crystallized.

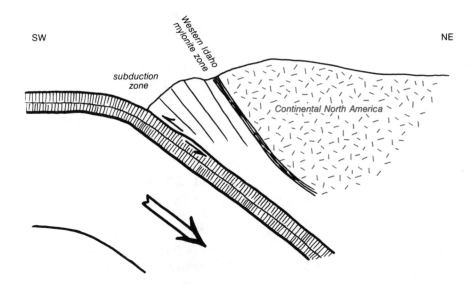

How the sinking slab of oceanic crust sheared the rocks in the continent above to create the mylonite zone near the western edge of Idaho.

Strongly sheared mylonite near the Dworshak Dam north of Orofino.

The sheared mylonites evidently formed just above the sinking oceanic crust, right where it slid beneath the old western edge of the continent. Rocks from one end of the mylonite shear zone to the other contain a very strong linear fabric. Near Orofino that lineation consistently points almost directly northeast. That, it seems, was the direction of the shearing, the direction the sinking slab was moving. If so, then the wandering islands came in from the southwest.

THE BITTERROOT AND ATLANTA BATHOLITHS

a.k.a THE IDAHO BATHOLITH

Earlier geologic maps of Idaho lumped granite of all ages together with basement rock. The result was a map that showed a vast blotch of unrelieved pink stretching across most of central Idaho, all supposedly uniform granite. Geologists called that great expanse of granite the Idaho batholith, a term

The Idaho batholith consists of two large expanses of granite, the Bitterroot and Atlanta batholiths. Both formed during late Cretaceous time. Smaller areas shown in the finer pattern are masses of Eocene granite.

still in widespread use even though the granite is neither so uniform nor so continuous as it once seemed.

Modern geologic maps show a far more interesting picture: two large areas of Cretaceous granite with the much older metamorphic basement rocks along the Salmon River between them and areas of metamorphic rocks around their margins. Smaller masses of much younger granite emplaced during Eocene time, about 50 million years ago, spot the map. We call the large expanse of Cretaceous granite north of the Salmon River the Bitterroot batholith, the southern mass the Atlanta batholith. Each intrusion of Eocene granite has its own name. But old traditions die hard.

Like most geologists, we still say Idaho batholith when we mean to refer vaguely and collectively to all that Cretaceous granite in central Idaho. It isn't always worthwhile to pick into the distinctions; loose and inexact terms have their uses.

Granite, Gneiss, and Schist

Below a depth of about 15 miles, the heat of all that rising basalt magma melted much of the rock in the continental crust to form granite magma. Somewhat cooler rocks above that depth merely recrystallized into metamorphic rocks — gneiss and schist. The granite magma rose because it was light, as though it were an enormous bubble, dragging its surrounding metamorphic rocks along. What you see now is granite set within broad zones of gneiss and schist.

Minerals in the metamorphic rocks tell something of the temperature and pressure that prevailed where the rock recrystallized, and that in turn sheds some light on the granites they enclose. To judge from the testimony of the metamorphic minerals, the magma that became the Bitterroot batholith rose to a depth of about ten miles, that of the Atlanta batholith about seven miles, before they crystallized into granite. The crystallization depth depends upon the amount of water in the magma.

This streaky outcrop is an exposure of mixed gneiss and granite — what happens when a metamorphic rock gets so hot that it begins to melt.
— U.S. Geological Survey photo by C.P. Ross

Rocks in the upper part of the continental crust and the lower part of the Precambrian sedimentary pile do contain some water, most of it incorporated in minerals such as mica. All that red hot steam rising from the sinking slab of oceanic crust into the continental rocks added more water. The water reduced the melting temperature of the rocks, making it much easier for the granite magma to form, meanwhile ensuring that the magma would be rich in steam. Such magmas do strange things.

A heavy charge of steam causes the melting temperature of a magma to rise with decreasing pressure — exactly the reverse of normal behavior. So, the steamy magmas of late Cretaceous time rose through the earth's crust, their melting point steadily rising as they ascended into regions of lesser pressure. Finally, they reached a level where the melting point matched the actual temperature of the magma. There the magma crystallized into solid granite, which then cooled at leisure. It is the drop in pressure, not a drop in temperature, that causes huge masses of granite to crystallize.

Seven or ten miles deep, either one, is a long way down, a lot of rock to remove before the sun could shine on the granites of central Idaho. So it is surprising to find that the Challis volcanic rocks that erupted 50 million years ago buried hills eroded into that granite. It is almost beyond imagining to suppose that the slow processes of erosion could have stripped off so much extremely resistant rock so quickly. Certainly not, considering that the Challis volcanic rocks that erupted 50 million years ago still survive, even though they were never more than a few thousand feet thick, and are much softer than the ten miles of rock that originally covered the granite.

Another zone of mylonite, profoundly sheared granite, several thousand feet thick lies along the east face of the Bitterroot Mountains in western Montana. It tells part of the story of what happened to the rocks that once covered the northern part of the Idaho batholith. In this case, the shearing appears to have happened as the cover of Precambrian sedimentary rocks moved east off the Bitterroot batholith into western Montana, where it forms several mountain ranges.

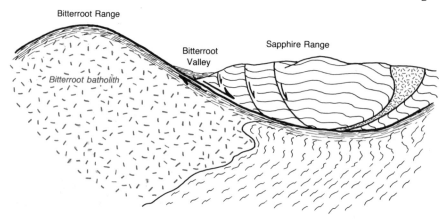

How the cover of the Bitterroot batholith moved east into western Montana, exposing the granite almost as it crystallized.

The eastward movement began while a large volume of molten magma still lurked within the early mountainous welt, making it as weak inside as an aging cream puff on a summer afternoon. Central Idaho must have been higher then than it now is, and the rocks above that weak magma moved east, down the regional slope. Removal of those seven or ten miles of covering rock greatly reduced the pressure on the granite magma, causing any of it that was still molten to crystallize immediately.

It is a puzzle that no corresponding mylonite has been found east of the Atlanta batholith. However, several mountain ranges there do contain large faults, which clearly carried the rock above them long distances to the east, almost certainly uncovering the Atlanta batholith. Mylonite forms deep in the crust where the rocks are so hot and therefore so soft that they flow plastically, like warm wax. Perhaps the seven or so mile depth at which the Atlanta batholith crystallized was not quite deep enough to permit the rocks to flow. They broke along faults, instead.

Strongly crumpled sedimentary rocks east of the Atlanta batholith.
— U.S. Geological Survey photo by C.P. Ross

THE EOCENE SCENE: A SECOND
GENERATION OF IGNEOUS ROCKS

A second episode of widespread igneous activity happened during Eocene time, about 50 million years ago. It added a good many granite intrusions to central Idaho, injected thousands of dikes, and produced the enormous volcanic piles of the Challis formation. Intrusions of Eocene granite scatter widely through central Idaho, and straggle east into central Montana. Volcanic activity concentrated along a broad zone that trends northeast from the Boise area into central Montana. The dikes follow that same zone.

All that widespread igneous activity is hard to explain. By the time it happened, the plate collision off the west coast had long since shifted west to a line that trends through the Okanogan Valley of British Columbia and north-central Washington, then southwestward through Oregon. The line of that old trench is much too far west to explain all the Eocene igneous rocks in central Idaho and Montana.

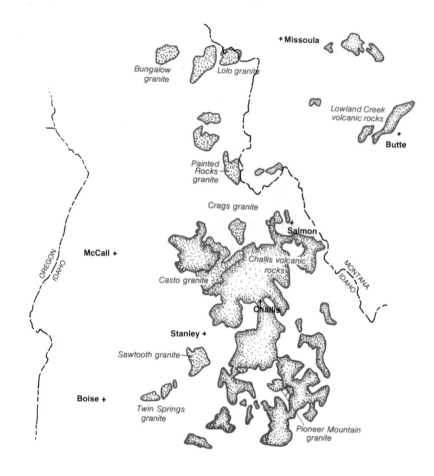

The distribution of Eocene granite and volcanic rocks.

Furthermore, a perfectly normal chain of Eocene volcanic rocks exists east of that former trench, showing that the sinking oceanic crust was descending at about the usual angle, and melting the crust above it at about the typical distance from the plate boundary. The kinds of rocks pose another problem. The Eocene volcanic rocks of central Idaho and western Montana are andesite in the lower part, rhyolite in the upper part; volcanic chains associated with a trench normally erupt large quantities of andesite. Those serious problems of location and composition make it dificult to argue that the widely scattered Eocene igneous rocks east of central Idaho and in western and central Montana owe their existence to a slab of oceanic crust sinking at an abnormally flat angle.

Horizontal cordwood jointing in a vertical dike cutting Challis volcanic rocks, just south of Challis.

In fact, the explanation of the Eocene igneous activity may require at least two ideas. Most of the volcanic centers and granite intrusions lie right along the trend of the Eocene dike swarm. Others lie along a crude arc that seems to mark the edge of one of the big slabs that moved off the Bitterroot batholith and into western Montana.

The Challis volcanic rocks of central Idaho erupted where the dike swarm cuts across the Atlanta batholith. Follow the trend of that swarm northeast to the Boulder batholith of western Montana, and it leads directly into the Lowland Creek volcanic pile. Both of those large volcanic areas also contain intrusions of Eocene granite.

The Atlanta and Boulder batholiths were only about 20 million years old when the dike swarm cut across them. They were still fairly hot inside. As the earth's crust stretched to open the cracks that filled with magma to become dikes, that relieved pressure on the hot granite below, allowing it to begin melting.

The arc of granite intrusions and rhyolite volcanic centers that seems to mark the outline of one of the big slabs that moved off the Bitterroot batholith may also owe its origin to pressure-relief melting. Some age dates on the Bitterroot mylonite in western Montana suggest that it may still have been moving in

Big chunks of rhyolite float in a matrix of rhyolite ash in this volcanic mudflow near Mackay, part of the Challis volcanic pile.
—U.S.Geological Survey photo by C. P. Ross

Eocene time, still unloading the Bitterroot batholith. The final unloading may have caused pressure-relief melting within the Cretaceous granite, which was still hot then.

Dikes, Granite, and Rhyolite

The great swarm of Eocene dikes trends northeast from the Boise area through central Idaho and southwestern Montana to the vicinity of Butte. They vary in composition. Some of the rocks are dark, basalts from the earth's mantle; others are pale rocks generally similar to the rhyolites of the Challis volcanic pile and the larger granite intrusions. Pressure-relief melting may be the origin of both the dike rocks and the Challis volcanic rocks.

With a bit of practice, you can learn to recognize the Eocene granites. Freshly broken surfaces are generally pale pink, whereas those on the older granites are pale gray. All of the Eocene granites are perfectly massive and the mineral grains tend to be distinctly larger than in the older granites.

Unlike the earlier magmas that became the Idaho batholith, those of Eocene time either erupted into sheets of rhyolite ash,

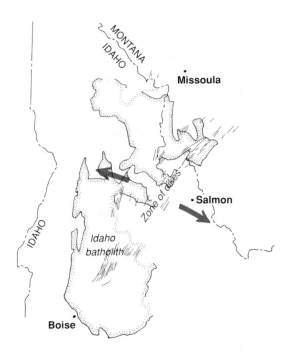

The zone of Eocene dikes, with arrows to show the direction of crustal stretching that must have opened the fractures they fill.

or crystallized into granite at extremely shallow depth. Evidently, the younger magmas contained very little water when they melted, and rose almost to the surface before they absorbed much. That is hardly surprising; the young granites of the Idaho batholith that melted to form those younger magmas probably contained virtually no water.

Everything that could happen to granite magma did happen as the Challis volcanic pile formed. Big rhyolite explosions continue for hours as escaping steam blasts shreds of molten magma out of the volcano. Part of the magma blows high in the air, then drifts in the wind as a plume of volcanic ash that slowly settles to the ground, like snow. Most of it boils out of the volcano in a dense cloud of molten ash and red hot steam that races across the countryside, an ash flow. Large parts of most ash flows are still so hot when they finally come to rest that the shreds of magma weld themselves into solid rock. Other masses of magma that happened to reach the surface before they

Intricately eroded rhyolite ash near Bayhorse in central Idaho.
— U.S. Geological Survey photo by C.P. Ross

absorbed much water became big rhyolite plug domes and lava flows. And the pile contains large granite intrusions that crystallized just before they broke the surface.

The upper part of a mass of dry magma may push above the surface to form a plug dome, while the lower part crystallizes below the surface as an intrusion. Rocks within both generally consist of fairly large crystals of feldspar and quartz set in an extremely fine-grained matrix. They look sort of like granite, and sort of like rhyolite. Think of them as rhyolite that almost became granite, or as granite that narrowly escaped being rhyolite.

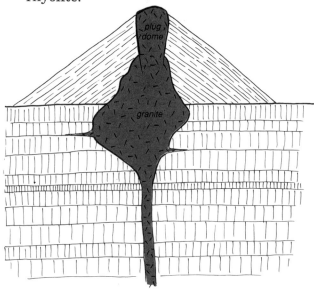

A rhyolite plug dome may be the top of a shallow granite intrusion.

110

Volcanoes erode even as they form. Rainwater washes sediment down their slopes. Big mudflows form where water soaks large volumes of ash, then pours down the slope, picking up larger rocks on the way. The sediments accumulate in valleys low on the slopes, locally in lakes that fill craters or valleys dammed by lava flows or mudflows. Eroded and redeposited volcanic rocks of one kind or another comprise a large proportion of most volcanic piles, including the Challis volcanic rocks.

COLUMBIA PLATEAU AND
THE BASIN AND RANGE

Central Idaho lies just east of the Columbia Plateau, just north of the Snake River Plain. Both started with the explosion of a giant meteorite in southeastern Oregon about 17 million years ago. Imagine the giant meteorite suddenly appearing as a dark mass in the sky, then burning its way through the atmosphere in a quick flash of fast running light that for just an instant burned brighter than the Sun itself. Then the blinding flash of the explosion, and the long rumbling as dark clouds of broken debris shot across the ground surface beneath an ominous mushroom cloud rising and spreading through the atmosphere. Earthquakes of extraordinary violence must have reverberated for hours.

The long aftermath of that bad day continued for several million years. It added some interesting rocks to central Idaho.

Flood Basalts and Tropical Lakes

By late Miocene time when the great meteorite struck, central Idaho had developed a landscape that must have looked very much like what we now see, streams flowing through most of the same valleys our modern streams follow. If you could magically transport yourself back to the time just before the great impact, we believe you would easily recognize most of the landscape of central Idaho. But the landscape of the lowlands that stretched away to the west would soon change beyond all recognition.

Probably within a few thousand years after the impact, the big crater in southeastern Oregon filled to the brim with molten basalt, and began to overflow. Great floods of basalt lava spilled across those western lowlands to begin building the Columbia Plateau. Some of the flows flooded areas the size of the state of Maine within a few days, no doubt with enormous environmental effects.

Imagine the effect that such an enormous area of red hot rock would have on the atmosphere, the violent storms it might cause. And consider that such a lava flow would release untold thousands of tons of carbon dioxide and sulfur oxides into the air. The carbon dioxide might well create a greenhouse effect, perhaps warming the climate of the entire planet. The sulfur oxides would react with atmospheric water to cause acid rains, probably very acid rains. Meanwhile, that lava lake in southeastern Oregon was blowing more gases into the atmosphere.

The eastern edges of the basalt floods backed into the stream valleys of central Idaho, impounding lakes. It is easy to imagine that some of those lakes must have been exceptionally beautiful in their mountainous settings, especially in the years between flood basalt flows. We can only speculate about the exotic birds and butterflies that flitted around their shores, but you can actually see some of the leaves that grew on the trees.

Streams washed layers of sand and gravel into those lakes and occasional rhyolite eruptions in southeastern Oregon sent clouds of pale volcanic ash drifting across, leaving layers of

White layers of Latah formation beneath a basalt flow in a road cut east of Lewiston.

A mass of basalt pillows and pale brown palagonite formed as a basalt flow boiled into a late Miocene lake near Coeur D'Alene.

white ash in their wake. Eventually, perhaps after thousands of years, a later flood of lava poured, steaming and hissing, into the lake. Molten basalt pouring into cold water breaks up into rounded masses, pillows. You often see them set in a matrix of yellowish clay called palagonite that formed as hot basalt ash reacted with water. Underneath the pillows and the palagonite are the lake sediments, the Latah formation. Watch for their white layers sandwiched between lava flows.

People who split layers of white rhyolite ash in the Latah formation are likely to find black impressions of fossil leaves as beautifully preserved as though they had been lovingly pressed between the pages of a dictionary. Many are easy to recognize as the leaves of trees that thrive today in the southeastern United States and on Caribbean islands. They leave us in no doubt that the climate of Idaho was indeed very warm and wet while the flood basalts of the Columbia Plateau erupted.

Indeed, the climate of the Northern Rockies was warm and wet enough to weather its rocks into red laterite soils like those typical of wet tropical and subtropical regions. Those red soils washed off the mountains and into the lakes, where they became the snow white sediments of the Latah formation. That transformation from red soil to white sediment requires some chemical magic.

Decaying vegetation can make lake waters acidic and reducing enough to dissolve the red iron oxide that stains laterite soils, leaving deposits of pure white clay, kaolinite. We suspect that acid rains following the enormous volcanic eruptions that built the Columbia Plateau may have helped bleach the lake sediments. In any event, the same acidic environment that bleached all the red iron oxide out of the clay would also dissolve bones. If fish swam in those lakes, their bones disappeared without a trace. So did the remains of any animals that may have been buried in the lake sediments.

Basin and Range

Basin and Range faults began to move as the Columbia Plateau began to form, almost immediately after the impact of 17 million years ago. They first broke the continental crust in northern Nevada and western Idaho, because that part of the state was closest to the site of the event. Now, the effects of those movements are most obvious in the eastern part of central Idaho, where they continue to move to the rumbling accompaniment of occasional earthquakes.

The fault block basins and ranges of east central Idaho.

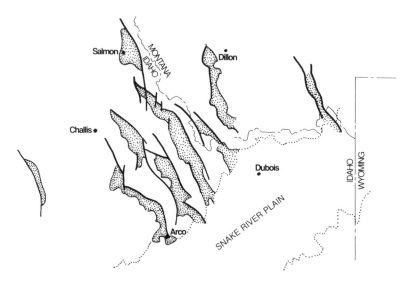

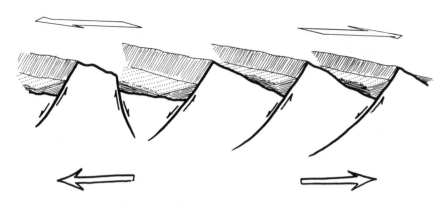

How movement on the Basin and Range faults dissects the Rocky Mountains into blocks that move away from each other.

Most of the Basin and Range faults trend northwest, so movement along them must stretch the earth's crust from northeast to southwest. The Basin and Range faults of central Idaho chopped the Rocky Mountains into blocks that pull away from each other while some rise to become mountain ranges and others drop to become valleys. As the crustal stretching continues, the ranges draw farther apart and the valleys between them grow wider. Meanwhile, rhyolite and basalt magmas rising along the Basin and Range faults erupt to form the youngest volcanic rocks in central Idaho.

FRONTIER MINING

All the igneous activity, especially that of 70 to 80 million years ago, provided central Idaho with an impressive variety of metallic ore deposits. They chiefly tend to concentrate around the margins of the Bitterroot batholith, but the Atlanta batholith has its share, as do some of the granites that arrived 50 million years ago.

Prospectors began to trickle into the region to look for gold and silver during the Civil War years, flooded in a decade later. They did indeed find quite an abundance of gold and silver, along with an amazing variety of other metals, some of which did not become valuable for many years. Central Idaho owes much of its history, and modern character, to those ore deposits, and to the people who found and worked them. Their task was neither easy nor especially rewarding.

Placer Mining

Every discovery of a few flakes of gold in a stream bed inspired a frenzied stampede of aspiring miners. Those must have been desperate times if a territory as thinly settled as central Idaho was a century or more ago could so quickly produce such large crowds of apparently unemployed men. Most of them found little opportunity in the mining camps, either.

Starting a small mine has always been an expensive and risky venture, no less so in 1865 than today. Those early placer miners were hardly the free spirits we tend to imagine, now that their troubled times are history. They had to invest quite a lot of money and effort to pack themselves and their gear into a new mining district before they could locate a claim. And they had to be quick. Miners who arrived in a new district more than a few days after the big rush began were just in time to find all the choice ground staked; themselves relegated to working for someone who had arrived on time.

After he had staked a claim, the miner had to clear it of trees and brush, and in most cases strip at least several feet of essentially barren gravel off the surface to reach the richer gravel below. Then it was time to hand saw lumber to build sluice boxes, rockers, and other equipment. Most miners had to dig ditches to carry water to their claims. When all the preliminary work was done and the spring runoff filled the creek, the miner could start washing gravel through the sluice boxes to see how much gold his claim might hold. By that time, he had invested heavily in equipment and months of labor — his own and that of less fortunate miners who worked for him.

A few miners struck it rich; a few people strike it rich in Las Vegas today. Most miners considered themselves very lucky indeed to recover their investment in equipment and labor, plus a ten percent profit. It is extremely difficult to find accounts of large fortunes won by washing gold out of stream gravels.

116

Despite all the talk of sudden bonanzas, and there were a few, recovery of an ounce or two of gold per day from a placer claim was generally considered a pretty good haul. Two good summer seasons of honest work were enough to skim the best hand workable gravel from most placer claims. Then it was time for the patient Chinese to take over, many years later the big dredges. That is why so many of the early gold rush towns boomed for about two years, then went into steep decline.

Gold mining has historically been an industry that thrived best when times were hard, when large numbers of men could not find regular employment. Even in the busiest placer mining camps, as many as two-thirds of the men who came seeking a fast fortune were not only too late to stake a claim, but unable to find work of any kind. Many of those unfortunates nearly starved if winter caught them in a remote mining camp. At the least, they were suffering from scurvy long before Spring arrived. Nevertheless, the hypnotic allure of gold with its false promise of free money for the digging continued to draw thousands of men from one remote creek and ramshackle mining camp to another. Their next best hope was hard rock lode mining, underground.

Hardrock Mining

Although most early miners assumed that discovery of the bedrock source of gold in the perennially famous mother lode would make them rich, the reverse was more often true. Think of streams as natural sluice boxes; they concentrate heavy minerals so effectively that placer deposits are almost invariably more rewarding than their bedrock source. The main exceptions are in districts in which gold occurs within some mineral, instead of as the native metal.

Regardless of their skills at finding gold, and at drilling and blasting hard rock, very few of the early operators knew how to manage the business aspects of underground mining. In fact, a review of early mining history in the Rocky Mountains reveals very little evidence of economically shrewd management at any level. Most of the operators were simply incompetent as business men.

Early bedrock mining districts generally began by milling their ore with arrastras, crude grinding contraptions that used water or horse power to drag a large rock around and around over ore spread on a circular pavement of rocks. The dragging rock crushed the ore fine enough to free the particles of gold. Then it was a simple matter to wash the gold and other heavy metallic minerals out of the powder.

Arrastras had all the great advantages of cheapness. Any country blacksmith could build one right on the spot with some big rocks and a few scraps of iron and lengths of chain. Hundreds, if not thousands, of them must have worked the ores of central Idaho during the last century. You rarely see their ruins because nearly every one was torn up to recover and wash the ground ore that sifted into the cracks in the stone pavement. All that remains are litters of stones with flat faces ground on them, and a few lengths of rusting chain.

Arrastras had the great disadvantage of slowness. One of those contraptions could grind only a few tons of ore a day, no way to get rich quick. Besides, most mines could easily produce more ore than they could grind with arrastras, an impossibly frustrating limitation to operators whose main goal was to get rich quick. Even so, attempts to expand from arrastras to larger, faster, and more efficient stamp mills ruined many early mines.

Stamp mills do work much better than arrastras, but they are enormously more expensive and far too complex to build on the spot in a nineteenth century mining camp deep in the mountains of central Idaho. The early mining companies had to buy the machinery for their stamp mills from eastern factories, then ship it out to the Rocky Mountains. The cost per pound of hauling by wagon and pack train was astronomical. Next, they had to build sheds to house the mills, and dig ditches to provide them with water.

In many cases, the operators who had so recently been frustrated by the slowness of arrastras found that they could not mine ore fast enough to keep the expensive new stamp mill busy. They needed to expand the mine as they invested in mills that could grind more ore in an hour than the old arrasta could in a day. Many of their development schemes involved long tunnels and deep shafts to reach ore that in too many cases existed only in totally unfounded optimism. All too commonly, the main result of those big development schemes was conclusive proof that the ore body would never pay for the mill. Then continued operation of the mine could only postpone the inevitable bankruptcy. Careful estimation of ore reserves and the amount of capital investment they could profitably amortize was not one of the management techniques generally used in nineteenth century western mining.

Nevertheless, the early miners did produce quite a lot of gold and silver, even if not very often at a profit. It is extremely difficult to estimate the production of early gold mines. The government did not require record keeping until 1893, and even then the records did not become perfectly reliable. You can be sure that the announced returns of many early mines were intended more for promoting stock than truthfully recording the cold facts for the edification of a curious posterity.

Regardless of their accuracy, remember that total production figures for mines represent gross income, cash flow, which may or may not have included any profit. Most of those figures pale in comparison to the total cash flow of an average modern supermarket. In the end, the overwhelming majority of nineteenth century gold mines turned out to be non-profit organizations. Some of the most profitable operations did very little actual mining.

Scams

Many of those old mines and claims back in the hills were simply stock promotions. In a word, scams. The promoters would set up a few small buildings, buy some nice ore samples, and print a batch of colorful stock certificates. Then they went to one of the big eastern cities, sold the stock, and returned to the Rocky Mountains to enjoy the proceeds. It was an efficient system: The investors promptly lost their money without any intermediate fuss over operating a mine. If you run across what looks like an abandoned mine, but can't find much sign of spoil heaps, ore minerals, ditches, or haulage roads, you may well be looking at the relics of an old stock scam. An amazing number of so-called mines never shipped any ore, or even had any mineralized rock to mine.

Especially beware of those glowing accounts of fabulously rich gold claims, now mysteriously lost. For sure, no one ever lost a claim that had any gold in it. Nevertheless, every district that ever produced any bonanza ore from underground mines has its legend of a lost claim. That has more to do with the tendency of bonanza ore to find its way into empty lunch boxes than with any difficulty about remembering the locations of claims.

Stolen bonanza ore is as hot as any other kind of stolen property. The person who has it needs to get rid of it as soon as possible, most conveniently by fencing it. Every district that produced much bonanza ore had a local fencing operation that hid behind a cover story of a secret claim. From time to time, the fence would announce that he was on his way out to work his secret claim, then spend the evening going from one miner's house to the next buying up the stolen ore stashed under the back steps. Then he would sell the ore to the mill, claiming that it had come from his secret claim. His ore invariably looked suspiciously like that from the producing mines. There is absolutely no point in looking for one of those supposedly lost claims, now that the mines that produced its ore are long closed.

Cyanide

Many gold districts enjoyed a revival after the development of cyanide mills early in this century. The cyanide process recovers gold chemically, by dissolving it out of the rock. Cyanide mills are so efficient that they can extract gold profitably from ore much too lean to feed earlier types of mills, and from ores in which gold exists as an impurity in other minerals, rather than as the native metal. The new process made it possible to mine ore that would previously have been considered common rock. Many mines ran the spoil heaps left from their earlier operations through the new mills to extract the last traces of gold that the old mills had missed. That sort of secondary recovery still continues in some of the old mining districts, when the price of gold is high.

THE ICE AGES

As elsewhere in the northern Rockies, the landscape of central Idaho clearly shows the effects of two major episodes of glaciation. Everywhere, the younger moraines and other glacial features nestle within the much larger ones left by the much larger earlier glaciers. The younger glacial moraines have very little soil on them; the boulders that litter their surfaces are almost perfectly fresh.

Be careful, though, not to exaggerate the extent of ice age glaciation in central Idaho. Only the higher mountains, the ranges that make their own weather today, caught enough snow to support glaciers. Recognize those ranges by their jagged skylines and sparkling high mountain lakes. Gnarled peaks drop off into high cirques that look like they were scooped out of the top of the mountain. The glaciers that nestled in those cirques gouged the valleys below into deep troughs, then left the valley floor littered with boulders as they retreated.

No glaciers ever bit into most of the mountain valleys of central Idaho. At lower elevations, most of the region looks about as it would had the ice ages never happened. But many of the higher ridges do show the effects of cold climates, even if not the mark of the ice.

The Sawtooth Range, glacially sculptured granite, reflected in Stanley Lake. — U.S. Geological Survey photo by T.H. Kiilsgaard

Cold Climate Erosion

Many of the higher ridges of central Idaho have a lean and bony aspect, the look of many bedrock outcrops along the crest of the divide. Although never glaciated, those ridges are almost certainly the victims of ice age erosion through a process called soil flowage.

Frozen ground is water tight, automatically self sealing; any water trickling through it plugs the channel it was following as it freezes. In regions where the subsoil is permanently frozen, the surface soil thaws and collects rain and snow melt water during the warmer months until it becomes impossibly sloppy. You see that happen now on a small scale in the muddy weeks of early spring, when the ground stays sloppy until the frozen subsoil finally thaws, permitting the trapped water to drain out of the surface soil.

It seems quite likely that the subsoil on the higher ridges of central Idaho was permanently frozen during the ice ages. If so, the surface soil was saturated during all the months it was thawed, probably from April until October. The sloppy surface soil poured like thick porridge off the high ridge crests and down the slopes during the ice age summers.

Now you see rock outcrops on the soil stripped crests of those ridges, much deeper than normal soil cover on their lower slopes. Where it is exposed in a hole or ditch, that transported soil on the lower slopes looks like it had been stirred, almost a marble cake effect. For some reason that we can't begin to explain, those lower slopes often sound hollow if you stamp on them — never mind the sound, the ground is perfectly solid.

Dikes of white granite cut diorite.

U.S. 12
Lewiston — Kooskia
71 miles

The highway follows the beautiful Clearwater River between Lewiston and Kooskia. The low route along the river confines the horizon within the valley walls, making it difficult to appreciate from the road that the river cut its valley through the almost level surface of the Columbia Plateau. Higher vantage points off the highway reveal that broad surface, and occasional hills of older rock that rise above it like islands.

All the rocks exposed along the road between Lewiston and the area a few miles west of Orofino are basalt lava flows of the Columbia Plateau, erupted about 16 million years ago. Between the area a few miles west of Orofino and Kamiah, the road alternately passes outcrops of basalt and older rocks, mostly medium gray rocks called diorite that resemble granite except in being darker and almost without quartz. Recognize the basalt by its dark rusty gray color and tendency to break into columns. The much older diorites are medium gray on fresh surfaces, various shades of darker gray or pinkish brown on weathered outcrops.

Diorite

Age dates show that the gray diorite exposed between Kamiah and Orofino is about 82 to 90 million years old, distinctly older than most of the Idaho batholith. Some geologists interpret the diorite as an

123

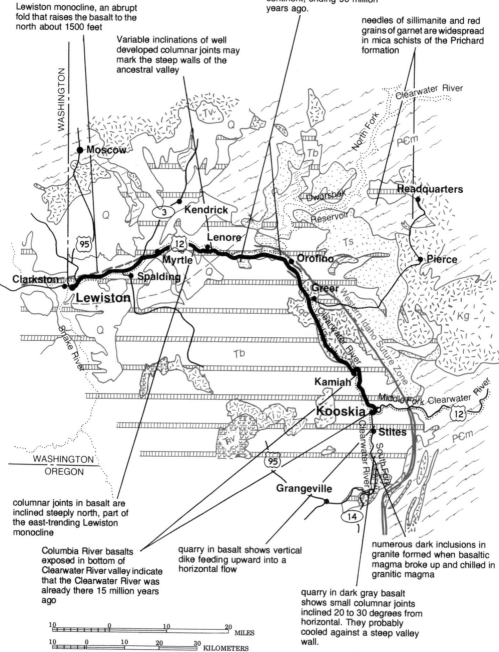

US 12
Lewiston—Kooskia

N

quartz diorite along US 12 and below 700-foot high Dworshak Dam is intensely sheared — the mylonite, marking collision of the Seven Devils island arc with the old North American continent, ending 90 million years ago.

needles of sillimanite and red grains of garnet are widespread in mica schists of the Prichard formation

Lewiston monocline, an abrupt fold that raises the basalt to the north about 1500 feet

Variable inclinations of well developed columnar joints may mark the steep walls of the ancestral valley

WASHINGTON

Clearwater River

● Moscow

Q

Tv

Q

Tb

North Fork

PCm

Dworshak

● Headquarters

Reservoir

95

3

Kendrick

12

Lenore

Ts

Orofino

● Pierce

Clarkston

Myrtle

Spalding

Q

Greer

Kqd

Western Idaho Suture Zone

Kg

Lewiston

Q

Clearwater River

Snake River

Tb

Kamiah

Middle Fork Clearwater River

Kooskia

12

WASHINGTON
OREGON

Tv

Ki

95

● Stites

South Fork

Clearwater River

PCm

columnar joints in basalt are inclined steeply north, part of the east-trending Lewiston monocline

Grangeville

14

Columbia River basalts exposed in bottom of Clearwater River valley indicate that the Clearwater River was already there 15 million years ago

quarry in basalt shows vertical dike feeding upward into a horizontal flow

numerous dark inclusions in granite formed when basaltic magma broke up and chilled in granitic magma

quarry in dark gray basalt shows small columnar joints inclined 20 to 30 degrees from horizontal. They probably cooled against a steep valley wall.

10   0   10   20 MILES
10 0 10 20 30 KILOMETERS

early part of the Idaho batholith. Most batholiths do appear to have started forming with emplacement of relatively small masses of dark rocks, then proceeded to much larger volumes of granite. The diorite rose into the suture zone along the old continental margin about 90 million years ago. The mineral composition of diorite with its shortage of quartz and abundance of dark minerals is more like that of oceanic than continental crust. Regardless of how you view the origin of the diorite, there seems little doubt that it lies along the old western margin of North America.

The Old Continental Margin

At Orofino, the highway passes through the old western margin of the North American continent, a geologic spectacular if ever there was one. Rocks west of Orofino are mostly diorite, those east of town are granites and granitic gneisses. The former continental margin appears in exposures of highly sheared diorite that form a zone about a mile thick just north of the Clearwater River. Watch for dark, streaky looking rocks, gneisses that tend to break into slabs, mylonites. The best places to look at those rocks are just north of Orofino, along the road beside the river to the base of Dworshak Dam, and at both ends of the dam. Gorgeous exposures of mylonitic diorite nearly surround the Visitor Center at the dam.

Geologists call rocks strongly deformed in shear zones mylonites. This mylonite began its career as massive diorite, essentially structureless rock. Then, shearing between the old continental margin and

Mylonitic diorite gneiss exposed in a road cut between Ahsaka and Dworshak Dam.

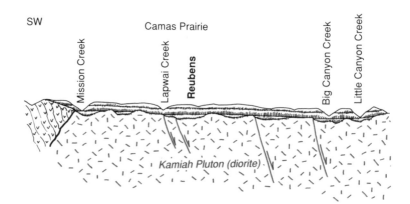

Section across Highway 12 near Orofino. The sheared mylonite in the northeast and oceanic islands that docked about 100 million years ago.

the sinking seafloor beneath smeared it into a platy gneiss that tends to break into slabs. Think of that shearing as resembling the kind of movement that happens when you slide a deck of cards, and think of the platy layers in the gneiss as cards in the deck. The displacement is distributed through the entire deck of cards, through the entire mile or so thickness of mylonite.

In this case, the shearing appears to have happened just above the slab of lithosphere that was sinking into the mantle, just west of the old continental margin. The main reason for coming to that conclusion is the position of the shear zone exactly where the continental granitic rocks give way westward to a more oceanic association of darker rocks that includes the diorite. That change in rock types almost certainly marks the old western margin of the continent.

Platy layers in the mylonite along the road to Dworshak Dam dip down to the northeast at about 55 degrees, probably somewhat steeper than the angle the sinking ocean floor slid down into the mantle, about 35 degrees. If you look closely at the surfaces of the layers, you will see that they have faint lines on them, a lineation. Geologists interpret those lines as a record of the direction of shearing — almost exactly northeast.

If the lineation points northeast, then apparently the sinking ocean floor was also heading northeast, coming from the southwest. That was the same sinking ocean floor that brought the rocks of the Seven Devils complex and Riggins group onto the old western margin of the continent. The tropical affinities of the fossils in the rocks of the Seven

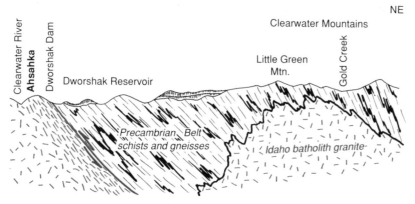

NE

Clearwater Mountains

Little Green Mtn.

Gold Creek

Clearwater River

Ahsahka

Dworshak Dam

Dworshak Reservoir

Precambrian Belt schists and gneisses

Idaho batholith granite

suture zone marks the boundary between the old continent to the

Devils complex fit with the evidence that they came from a southwest-erly direction.

Age dates show that the youngest mylonites are approximately 82 million years old. The absence of younger mylonites suggests that all shearing movement along the old continental margin ceased about 82 million years ago. Evidently, the former oceanic islands that became the Seven Devils and Riggins complexes of western Idaho were firmly fixed to North America by that time.

The Columbia Plateau

Just west of the junction with Idaho 3, the highway passes several large exposures of Columbia Plateau basalt that tilt steeply down to the south. This is near the eastern end of the Lewiston monocline, a sharp fold in the Columbia Plateau that flexes the north side sharply up. The basalt flows appear to drape over a fault that displaces the rocks at depth. A few miles farther east, the basalt flows tilt much less steeply in folds on the south side of the river, but are nearly horizontal on the north side. Evidently, the fold trends east essentially along the line of the river, and fades eastward.

The road passes the eastern edge of the Columbia Plateau in piecemeal fashion between Orofino and Kooskia. Watch for exposures of gray diorite of the old continental margin that lie under the dark plateau basalts with their characteristic columns. This part of the Clearwater River appears to have cut a new valley since the flows erupted.

The basalt flows flooded all the old river valleys in western Idaho to give the eastern margin of the Columbia Plateau a raggedly embayed map outline. In the Kooskia area, the flows apparently filled the old valleys completely, creating a new volcanic surface on which this part of the Clearwater River began to flow. Now, the modern river cuts across the old landscape, alternately passing through buried ridges of old continental rocks and old valleys filled with the much younger basalt lava flows.

The Kamiah Buttes

The Kamiah Buttes are a group of three high hills that rise some 1000 feet above the Columbia Plateau about eight miles west of Kooskia, the same distance southwest of Kamiah. They are conspicuous landmarks from the plateau surface; almost impossible to spot from the road in the canyon floor.

All three Kamiah Buttes consist of andesite and related volcanic rocks, so they must be the eroded remains of a volcanic center. It seems reasonable to assume that the volcanoes were probably active during Eocene time, that they are an isolated outpost of the widespread activity that produced the enormous Challis volcanic pile in central Idaho. Long after the volcanoes erupted for the last time, basalt lava flows of the Columbia Plateau flooded the countryside around them. Now the Kamiah Buttes stand above the plateau surface, lonely survivors of a landscape mostly buried under basalt.

The Winding Canyon

All along the route, the road follows the floor of a broadly winding canyon cut into the plateau basalt. It seems likely that the canyon inherited those bends from big meanders the Clearwater River

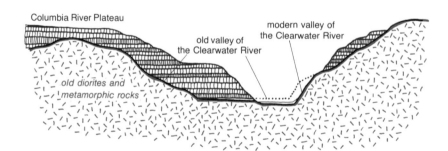

Schematic section across the Valley of the Clearwater River showing the old valley, the basalt that fills it, and the much smaller modern valley eroded into the basalt.

followed when it began flowing on the smooth surface of the Columbia Plateau. The river continued to follow those bends as it entrenched its valley deeply into the basalt.

The Clearwater River now flows on basalt along most of the way between Lewiston and Kooskia, so it still has not cut back down to the level of the valley floor that existed before the great flood basalts came this way. The modern valley eroded into the basalt is considerably narrower than the old valley eroded into granite, no doubt because the modern river is much smaller than the stream that flowed here before the Columbia Plateau basalts erupted.

Fibrolite

When the Clearwater River is low, rock hounds prowl the gravel bars between Lewiston and Orofino. They seek translucent pebbles of white or pale blue fibrolite, an uncommon variety of the common metamorphic mineral sillimanite. They polish the pebbles into iridescent little cabochons that shimmer in the light.

Sillimanite is an aluminum silicate mineral that occurs in schists that contain large amounts of aluminum and recrystallized at extremely high temperature. The North Fork of the Clearwater River drains a large area of such high grade schists, which explains why the good places to look for fibrolite pebbles are below Orofino, where the North Fork enters the main stream.

Sillimanite typically forms minute fibrous crystals, fine mineral hairs that tend to plaster the surfaces of mica flakes. They are easy to see with a microscope, barely visible to a practiced eye peering through a strong hand lens at a mica flake that looks slightly hairy. Fibrolite is simply a solid mass of sillimanite fibers. The silky look of polished specimens expresses their fibrous internal character.

US 12
Kooskia—Lolo Pass

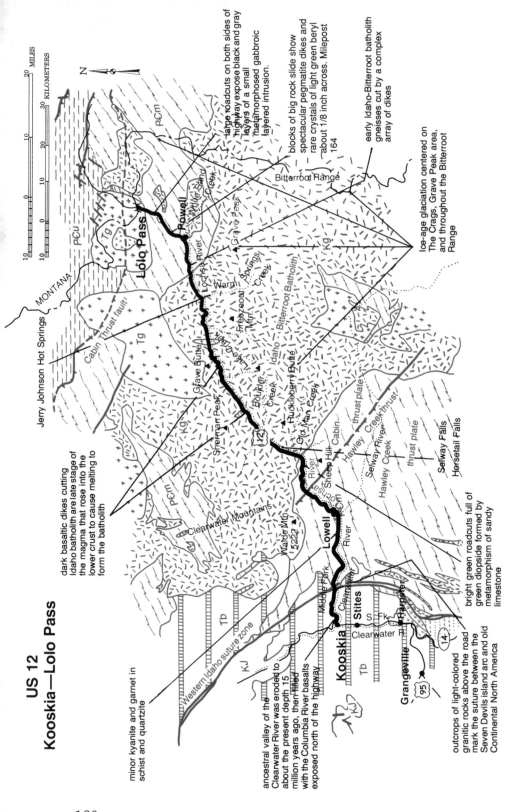

large roadcuts on both sides of highway expose black and gray layers of a small metamorphosed gabbroic layered intrusion.

blocks of big rock slide show spectacular pegmatite dikes and rare crystals of light green beryl about 1/8 inch across. Milepost 164

early Idaho-Bitterroot batholith gneisses cut by a complex array of dikes

Ice-age glaciation centered on The Crags, Grave Peak area, and throughout the Bitterroot Range

Selway Falls
Horsetail Falls

bright green roadcuts full of green diopside formed by metamorphism of sandy limestone

outcrops of light-colored granitic rocks above the road mark the suture between the Seven Devils island arc and old Continental North America

ancestral valley of the Clearwater River was eroded to about the present depth 15 million years ago, then filled with the Columbia River basalts exposed north of the highway

dark basaltic dikes cutting Idaho batholith are late stage of the magma that rose into the lower crust to cause melting to form the batholith

minor kyanite and garnet in schist and quartzite

Jerry Johnson Hot Springs

MONTANA

Cabin Thrust fault

Lolo Pass

Powell
White Sand Cr.
Lochsa River
Warm Springs Creek
Freezeout Mtn
Grave Peak
Bitterroot Range
Bitterroot Batholith
Idaho
Huckleberry Butte
Grave Butte
Sherman Peak
Boulder Creek
Old Man Creek
Sheep Hill Cabin
Lowell
Lochsa River
Hawley Creek thrust
Selway River
Hawley Creek
thrust plate
thrust plate
Clearwater Mountains
Walde Mtn 5222
Middle Fork
Clearwater River
S. Fk. Clearwater R.
Kooskia
Stites
Harpster
Grangeville
Western Idaho suture zone

N

MILES
KILOMETERS

130

Intersecting dikes of white granite in gneiss exposed in a roadcut near the Powell Ranger Station. The larger dike is about one foot across.

U.S. 12
Kooskia — Lolo Pass
101 miles

The route between Kooskia and Lolo Pass follows the Clearwater River to its source in the junction of the Selway and Lochsa Rivers at Lowell. Between Lowell and Lolo Pass, the road follows the Lochsa River to its headwaters on the Bitterroot Divide. The entire route is within sight and sound of a beautiful river, rarely out of sight of large road cuts and natural stream bank exposures of the bedrock. It is a lovely drive.

The route crosses the northern end of the Idaho batholith, as well as the broad zones of metamorphic rocks that embrace it on either side. The region offers no better place to get a good view of a fairly typical deep crustal batholith in its metamorphic setting.

The Bitterroot Batholith

For a distance of about 40 miles between the two metamorphic border zones, the highway crosses the Bitterroot batholith, the northern part of the Idaho batholith. Watch for road cuts in pale gray granite, mostly massive rock that tends to weather to a pinkish brown color on natural outcrops. But don't expect to see a vast expanse of

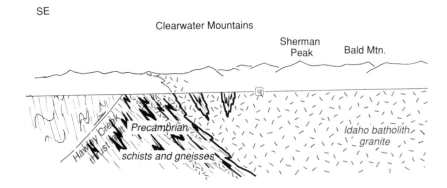

SE

Clearwater Mountains

Sherman Peak   Bald Mtn.

Precambrian

Hawley Creek thrust

schists and gneisses

Idaho batholith granite

Section drawn along Highway 12 between Lowell and Lolo Pass. Broad

absolutely uniform rock. Like most granite batholiths, this one is full of intriguing variations on the theme of granite.

Many road cuts near the borders of the batholith expose a peculiar granite full of oversized crystals of feldspar, slightly pinkish blocks set in a much darker matrix full of black biotite mica. In places, the blocky feldspar crystals are crudely aligned, probably by flow within the crystallizing magma. If you look closely at that rock, you will see that most of the large crystals of pink feldspar have white rims — they consist of orthoclase feldspar mantled by plagioclase feldspar. Geologists call such rocks Rapakivi granite, after the place in Finland where they were first studied. We have read many theories that attempt to explain their origin without finding one that is entirely convincing.

Crystals of feldspar almost the size of dominoes in granite exposed in a roadcut a few miles northeast of Lowell.

132

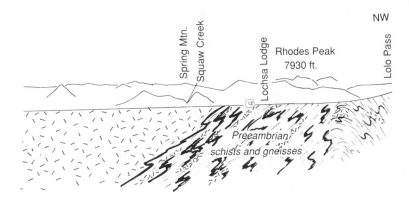

border zones of metamorphic rocks embrace the granite batholith.

Granite exposed in the road cuts along Highway 12 contains numerous veins of pegmatite an inch or two thick that lie almost flat. Watch for pinkish streaks of rock made of uncommonly large crystals an inch or more across. Many granites contain pegmatite, which appears to crystallize from the last watery remnants of the original magma. It is not clear why these in the Bitterroot batholith tend to be so nearly horizontal. We think it may have something to do with stresses set up within the still crystallizing granite as its roof sheared off and moved east into Montana.

The northeastern part of the Bitterroot batholith spreads out nearly flat over the metamorphic rocks in the area east and southeast of Lolo Pass, assuming the form of a sheet several thousand feet thick — relatively thin, as such things go. We think the magma probably moved east of the main mass of the batholith along a nearly flat thrust fault; the Hawley Creek or Cabin thrust faults seems likely candidates. If so, then it probably follows that the molten magma would have lubricated the thrust fault, which moved until the magma finally locked it tight by crystallizing into rigid granite.

Dark Dikes in the Granite

Watch the road cuts in the Idaho batholith, especially in its western half, for basalt and dark andesite dikes that look like more or less vertical stripes of dark rock in the pale gray granite. Most of the dikes meet the road at an angle that makes them more easily visible to people travelling east. The difficulty of seeing those dikes is truly

The complex pattern formed as molten basalt and molten granite stirred into each other in one of the dark dikes east of Lowell.

deceptive; actual measurements show that in some areas they account for something like 20 percent of the volume of rock.

Close examination of the dikes shows that many seem to mix into the lighter granite in rounded and streaky masses that convey much the same general impression you get while watching dark ink mix into clear water. Other exposures show granite injecting fractures in dikes exactly as the dark dike rock fills larger fractures in the granite. If the dikes inject the granite and the granite injects the dikes, then both kinds of magma must have been molten at the same time.

Remember that granitic magmas are extremely stiff and pasty, so the idea of a fracture opening in them is not so strange as it may seem at first thought. Consider also that the dark magma crystallized at a temperature hundreds of degrees higher than the melting point of granite. That makes it easy to imagine the dark magma injecting a fracture opening in the granite magma, then freezing to become a solid dike while the granite was still molten. Continued movement of the molten granite might then break the solid dike, opening fractures that the granite could inject to make dikes within dikes.

Dikes must form as magma injects an open fracture, and open fractures can form only when something stretches the rock. Obviously, the Idaho batholith stretched as it was emplaced. But the dark dikes do not, so far as we know, extend much beyond the granite into the older rocks surrounding it. Therefore, the stretching must have been limited to the batholith, could not have been regional. Furthermore, stretching must have happened while the granite was still at least partly molten. One way to sort all that out is to imagine the mass of granite magma spreading laterally as it rose, like rising bread dough spreading beyond the rim of a bowl. Such spreading would open fractures in the crystallizing granite magma that the darker magma of the dikes could fill.

The great abundance of dark dikes suggests that an enormous volume of basaltic magma must have existed somewhere deep in the continental crust while the granite magma of the Idaho batholith was rising and spreading into its present position. All that molten basalt may well have provided part of the heat to melt the granite magma that became the Idaho batholith.

Metamorphic Rocks —
The Border Zones of the Batholith

Almost all the road cuts between Kooskia and the western edge of the batholith, as well as those between the eastern edge of the batholith and Lolo Pass, expose metamorphic rocks. The only exceptions are occasional road cuts in granite, small outposts of the batholith. Recognize the metamorphic rocks through their general tendency to look layered, streaky, and even swirly.

Most of the metamorphic rocks are gneiss, rocks that resemble granite except in having a streaky appearance and containing bands of color. A close look reveals that the mineral grains in the metamorphic rocks tend to align parallel to each other — in military formation, so to speak. That disciplined alignment of mineral grains makes

Streaky granitic gneiss exposed in a roadcut near the Powell Ranger Station.

135

metamorphic rocks differ in appearance from one direction to another. See that difference by turning a specimen of gneiss or schist in your hand and watching its appearance change. Igneous rocks, on the other hand, are massive; they look the same from any direction because their mineral grains are randomly oriented.

The origin of the metamorphic rocks west of the Idaho batholith isn't completely clear. Most geologists contend that they began as Belt sedimentary formations. Others prefer to interpret the western metamorphic rocks as ancient continental basement that the rising granite magma dragged up. Both ideas are reasonable; neither rules out the other. The truth could well include a bit of both.

At least some of the metamorphic rocks east of the Idaho batholith must be metamorphosed Belt formations. Watch near Lolo Pass for road cuts in greenish rocks. Those are gneisses that contain an abundance of calcium minerals; they must have formed through metamorphism of a formation that contained limestone or dolomite. The obvious candidate is the Wallace formation, one of the Precambrian Belt formations. It has the same chemical composition as the green metamorphic rock near the top of the pass and even looks a bit similar. The Wallace formation appears in almost unmetamorphosed condition in road cuts just below Lolo Hot Springs, in Montana.

Many of the gneisses on both sides of the batholith weather to a rusty red color, an iron oxide stain that formed as pyrite weathered. Those rocks may well have formed through metamorphism of the Precambrian Prichard formation, the oldest of the Belt formations. Unmetamorphosed Prichard formation has the same chemical composition as the red-weathering gneisses, and it also contains pyrite.

All the metamorphic rocks within a few miles of the batholith are full of thin veins of pale granite. Most of those probably formed as

Scrunched dikes of pale granite in greenish calc-silicate gneiss near Lolo Pass. The largest dike is about two inches across.

magma injected between the layers of metamorphic rock. A few seem to have crystallized from magma that melted in place when the metamorphic temperatures got high enough to partially melt the rocks. That, after all, is how the magma formed to become the batholith.

The Lolo Batholith

Although several masses of younger granite exist near the highway, the Lolo batholith is the only one it actually crosses. Watch for the granite right at Lolo Pass. The best roadside exposures of this granite are a few miles farther east in Montana, around Lolo Hot Springs.

People who study the granites of central Idaho quickly learn to recognize those of 50 million years ago because they tend to consist of larger crystals than the older granites of the Idaho batholith, and the freshly broken rock is pink instead of gray. Those granites are also full of small gas cavities lined with very nicely formed crystals.

Most of those gas cavities are so small that you can't see them clearly without a good hand lens; a few are large enough to contain crystals an inch or more across. In some of the younger granites, including the Lolo batholith, those larger gas cavities contain gorgeous crystals of smoky quartz, real gemstones. People look for them along the logging roads after a hard rain.

The Peculiar Lochsa River

In long stretches of its lower part, the Lochsa River is almost as peculiar as it is beautiful. Notice that terraces, smooth remnants of old floodplains above the level of the modern floodplain, are very scarce and most inconspicuous. Most rivers have one or more prominent terraces along almost their entire length. Also notice that long stretches of the Lochsa River are almost continuous rapids, the lovely deep green pools much farther apart than you would expect. Most small rivers flow alternately through pools and rapids. And watch to see how minor tributaries enter the Lochsa River through steep cascades or waterfalls. That is most unusual in an unglaciated valley such as this. All those peculiarities suggest that the Lochsa River is rapidly entrenching its bed. Why?

No one actually knows why. One possibility might be that the land in this part of Idaho is rising, forcing the river to erode its channel to maintain its course close to the original level. That idea receives some support from gravity maps, which show that this part of Idaho has a slightly weaker than normal gravitational field. Such areas do tend to rise. To imagine why, think of what happens when you remove a load, such as a can of soda pop, from a floating air mattress. That reduces the mass of material riding on the water beneath where you removed the can, and the air mattress rises as water flows in beneath. The earth's crust behaves in much the same way.

Idaho's First Gold Rush

Pierce traces its origins back to late 1860, when early prospectors found gold in the Clearwater River and some of its tributaries. According to many accounts, this was the first gold discovery in Idaho. The usual crowd of frantic and incorrigibly optimistic miners, perhaps as many as 7000, stampeded into the area within a few months to take up claims, hoping to start mines. We never cease to marvel at the accounts of early mining rushes in the northern Rocky Mountains, the enormous numbers of apparently unemployed men who seem to have had nothing else to do but rush from the site of one rumor of gold to the next.

Several estimates place the annual production of placer gold at about a half million dollars until 1865 — in those days, an ounce of gold brought approximately 15 dollars. Those first few years skimmed off the easy pickings. Then Chinese miners took over, and patiently worked the streams for another 20 years, producing an unknown amount of gold. They must have picked the gravels to the bone; Pierce hasn't seen any placer mining worth mentioning since they left.

Hot Springs

The highway passes several hot springs, all most extraordinary in remaining nearly as nature made them. Where else do you see hot springs out in the woods without glowing signs, parking lots, concrete pools, and roadside bars?

Clouds of steam rise from the creek at Jerry Johnson Hot Springs, about a mile south of the highway.

It is tempting to assume that these, like most hot springs, produce water that was heated as it circulated to depths of a mile or two along deep fractures in the earth's crust. However, nothing on the geologic maps we have seen supports that assumption. No large faults that might open deep fractures are known to pass through these hot springs, nor do they lie along contacts between different rock formations. Satellite images show that Jerry Johnson Hot Springs lie on the margin of a big round feature, a target several miles in diameter. We don't know what it is.

Glaciation

Ice age glaciers were confined to the higher mountains in this part of Idaho; they never did reach low enough to get into the larger stream valleys. Watch for an occasional distant glimpse of a few craggy peaks, about the only sign of the ice ages anyone ever sees between Lolo Pass and Kooskia.

The approximate extent of glaciers in the Selway-Bitterroot area during the last ice age.

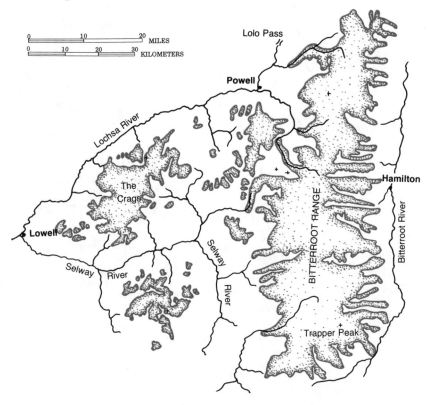

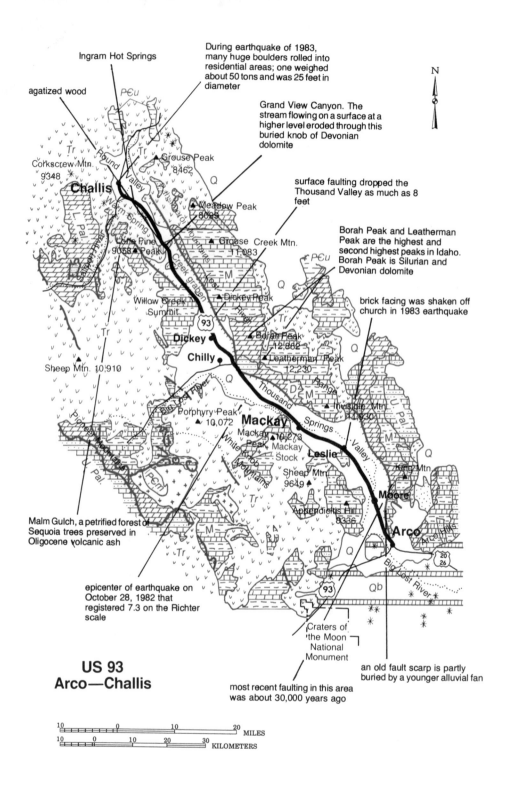

Ingram Hot Springs

agatized wood

During earthquake of 1983, many huge boulders rolled into residential areas; one weighed about 50 tons and was 25 feet in diameter

Grand View Canyon. The stream flowing on a surface at a higher level eroded through this buried knob of Devonian dolomite

surface faulting dropped the Thousand Valley as much as 8 feet

Borah Peak and Leatherman Peak are the highest and second highest peaks in Idaho. Borah Peak is Silurian and Devonian dolomite

brick facing was shaken off church in 1983 earthquake

N

p€u

Tr

Corkscrew Mtn. 9348

Tr

Challis

Grouse Peak 8462

Q

Meadow Peak 8095

Lone Pine Peak 9058

L Pal

W Fork Spring

Grouse Creek Mtn. 11,083

p€u

M

Q

Willow Creek graben

Willow Creek Summit

93

Dickey Peak

Tr

Dickey

Borah Peak 12,662

Q

Chilly

Leatherman Peak 12,230

Tr

Sheep Mtn. 10,910

Q

Big Lost River

Thousand

D

M

Range

Invisible Mtn. 11,930

U Pal

Pioneer Mountains

Q Pal

Porphyry Peak 10,072

Mackay

Mackay Peak 10,273

Mackay Stock

White Knob Mountains

Springs

Valley

M

Q

Leslie

Sheep Mtn 9649

King Mtn

Tr

Moore

Malm Gulch, a petrified forest of Sequoia trees preserved in Oligocene volcanic ash

M

Tr

Appendicitis Hill 8336

Q

Arco

Arco Hills

20 26

epicenter of earthquake on October 28, 1982 that registered 7.3 on the Richter scale

Big Lost River

Qb

93

Craters of the Moon National Monument

**US 93
Arco—Challis**

most recent faulting in this area was about 30,000 years ago

an old fault scarp is partly buried by a younger alluvial fan

10   0   10   20 MILES
10 0 10 20 30 KILOMETERS

140

Ledges of Paleozoic limestone in the hillsides near Arco.

US 93
Arco — Challis
78 miles

The route between Arco on the Snake River Plain, and Willow Creek Summit, follows the Big Lost River, so named because it disappears into the lava flows of the Snake River Plain. The Big Lost River flows through the floor of a broad valley, a Basin and Range fault block that drops as the adjacent ranges rise. Soaring mountains northeast of the road are the Lost River Range, a rising fault block that carries Borah Peak, the highest mountain in Idaho, on its crest. The much less impressive White Knob Mountains southeast of the highway are on another block.

Between Arco and the area near Mackay, the highway crosses basalt lava flows of the Snake River Plain, invisible beneath their deep blanket of stream sediments. The area near Willow Creek Summit is one of the few parts of the route where hard rocks are exposed along the road, not just in the distant hills. The pale gray outcrops there are exposures of the White Knob limestone, a very thick formation that tends to form ridges and cliffs. Fossils in the White Knob limestone show that it accumulated in shallow sea water during Mississippian time, about 350 million years ago. Most of the road between Challis, on the Salmon River, and Willow Creek summit follows a broad valley. Rocks exposed near that part of the road are rhyolite ash and soft basin-fill sediments.

141

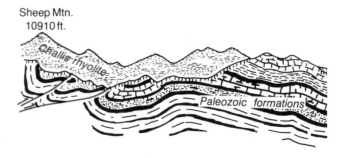

W

Sheep Mtn.
10910 ft.

Section across the line of the highway about midway between across the older structures in the mountains.

The Lost River Range and White Knob Mountains

While the Rocky Mountains were forming, great slabs of rock moved dozens of miles east on big thrust faults. Those displacements stacked slabs of older rock moving in from the west onto younger rocks that were already in this area. The moving slabs may have come off the top of the southern, Atlanta, portion of the Idaho batholith, sometime around 70 to 75 million years ago, perhaps while it was still partly molten.

The White Knob Mountains contain large granite intrusions that rose into them about 50 million years ago, penetrating the stack of thrust slices that had piled up while the northern Rocky Mountains formed. Large volumes of rhyolite, part of the Challis volcanic pile, that probably erupted from those granite intrusions deeply blanket much of the range, covering most of the older rocks.

Basin and Range faults that trend generally northwest broke the stack of overthrust slabs into blocks, raising some to form mountain ranges, dropping others to form valleys. The Lost River Range rises on the east side of one such fault, while the block on the west side drops and tilts. The White Knob Mountains rise on one side of that tilting block, while the valley drops on the other side. Meanwhile, the creeks keep the pace.

A side road leads six miles northeast of Leslie up the valley of Pass Creek to Bluejay Canyon, a spectacular deep gorge eroded through limestone in the Lost River Range. It seems clear that Pass Creek has eroded its bed fast enough to maintain itself at essentially its original

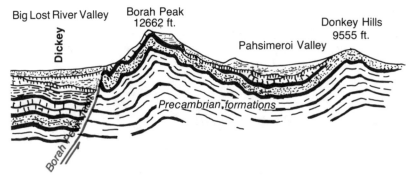

Arco and Challis. The young Basin and Range faults cut

level as the Big Lost River Range rose across its path. Many other stream valleys display exactly the same situation in less dramatic and less obvious terms. The big fault along its western face is still raising the Big Lost River range to the rumbling accompaniment of occasional severe earthquakes, such as that of October, 1983.

The Borah Peak Earthquake, 1983

Just after breakfast on October 28, 1983, the Big Lost River Range abruptly rose a foot or so while the adjacent valley dropped about four feet. Everyone in the Northern Rockies felt the shock. It was a Richter magnitude 7.3 earthquake, easily large enough to cause major devastation and loss of life had it struck a more heavily populated region. An earthquake of that size in Los Angeles or San Francisco would have caused a terrible disaster.

The little scarp that formed as the rocks on opposite sides of the fault moved in the instant of the earthquake is still plainly visible for several miles along the base of the Big Lost River Range. It is about six feet high along most of its length, a bit more near the center of the earthquake. Eyewitnesses saw it ripping along its length in a matter of seconds. Watch for that low cliff by looking east from the highway near the top of Willow Creek Summit, or take one of the unpaved roads east a mile or two for a closer view.

The earthquake shock waves destroyed several buildings in Challis, where falling masonry killed two children as they walked to school. The tremor snapped chimneys throughout a wide area that extended well into western Montana. Large boulders of rhyolite, including one

The low cliff in the foreground is the fault scarp that formed during the Borah Peak earthquake. The profile of the Big Lost River Range is on the skyline. — U.S. Geological Survey photo by H.E. Malde

that weighed about 50 tons, rolled down the mountain into Challis to join others that must have arrived during previous earthquakes.

Aside from the usual long series of rumbling aftershocks, the most striking after effects of the earthquake involved odd movements of ground water. For several days after the event, water gushed in fountains from the valley floor; for six months afterward, large springs flowed so copiously that the Big Lost River doubled its usual flow. Farther away, water flooded a silver mine at Clayton, while springs elsewhere dried up for a few days, then flowed more heavily than before. For a while, Old Faithful Geyser in Yellowstone Park erupted less frequently, and less faithfully.

Of course there is no way ever to know exactly what caused all those weird changes in ground water flow. Some of the fountains of water near the fault may have formed because the violent shaking of the earthquake compacted loose sediment in the valley floor, thus leaving less open pore space between sand grains and pebbles. Such an abrupt loss of pore space would force the water out of the rock. Fountains gushed from solid bedrock in the valley, because the sudden increase in elevation of the mountains raised the water pressure. Tilting of blocks of rock underground may well have set the water in them in motion, thus causing the more distant effects.

144

Just a few years before the big earthquake, members of the U.S. Geological Survey studied an old fault scarp about six feet high at the site of the Mount Borah earthquake. Their best estimates place its age at about 10,000 years. The new scarp of 1983 almost exactly duplicated the older one, so it seems reasonable to suppose that the earthquake of 1983 was an exact duplicate of one that happened about 10,000 years ago. A bit of arithmetic shows that about 1500 such earthquakes would be enough to explain the difference in elevation between the top of Mount Borah and the elevation of similar rocks buried beneath valley fill sediments on the opposite side of the Lost River fault. If we assume that the fault is about five million years old, that works out to a repetition of the Mount Borah earthquake every 3300 years, more or less.

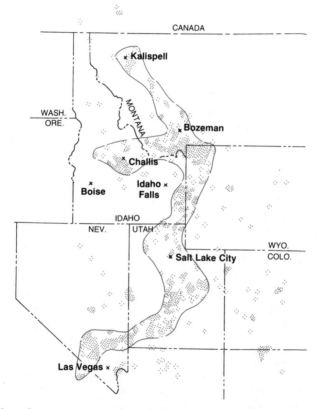

Earthquakes concentrate along the Intermountain Seismic Belt that forms an arc at the active eastern edge of the Basin and Range. A separate arm extends westward into central Idaho.

145

The road disappears into Grandview Canyon.

Grandview Canyon

A few miles south of Challis, the road abruptly passes through a narrow canyon eroded in dolomite of the Jefferson formation, which was deposited during Devonian time, about 400 million years ago. Even though the rock contains enough organic matter to stain it dark gray, it contains very few fossils. Perhaps the living organisms were mostly such things as algae, which rarely leave easily recognizable fossils. Or perhaps the water was too salty for the normal menagerie of small animals.

Why did the stream carve that narrow canyon in hard bedrock, instead of eroding a valley through the much softer sediments nearby? It had no choice. Grandview Canyon exists because the stream that eroded it began to flow on the old valley floor, above the level of the modern landscape, on a course that took it over the top of a completely buried hill. That probably happened about two or three million years ago, when the change to a wetter climate started most of the streams in this region.

As it eroded its channel, the stream came down on top of the buried hill, and then it had to cut that narrow canyon through the hard bedrock. It did not enjoy the alternative of escaping its valley to find an easier course over softer rock because that would require the stream to flow uphill. Streams never flow uphill.

146

Molybdenum

Challis, long a quiet little ranching community, may someday become a busy mining town. That almost did happen about 1980, when Challis enjoyed a brief boom after an enormous molybdenum deposit was discovered on Thompson Creek, about 40 miles southwest of town. Ambitious plans for an open pit mine that will cover almost a square mile and eventually reach a depth of nearly 2000 feet went into suspended animation after the price of molybdenum dropped from $23 a pound in 1970 to something less than $3 in 1985. Big mining will wait until economic conditions in the industry improve enough to make the deposit competitive on the world market, someday.

The Fizzled Mackay Copper Boom

Hardly anyone remembers that Mackay started out as a mining camp in a great surge of optimism, now long forgotten. A smelter was built there about the turn of the century to process copper ore from the White Knob Mountains west and southwest of town. But it soon became clear that the mine could not produce nearly enough ore to keep the smelter busy, and the project collapsed. The machinery was salvaged and shipped to California as Mackay lapsed into its long nap.

US 93
Challis—Lost Trail Pass

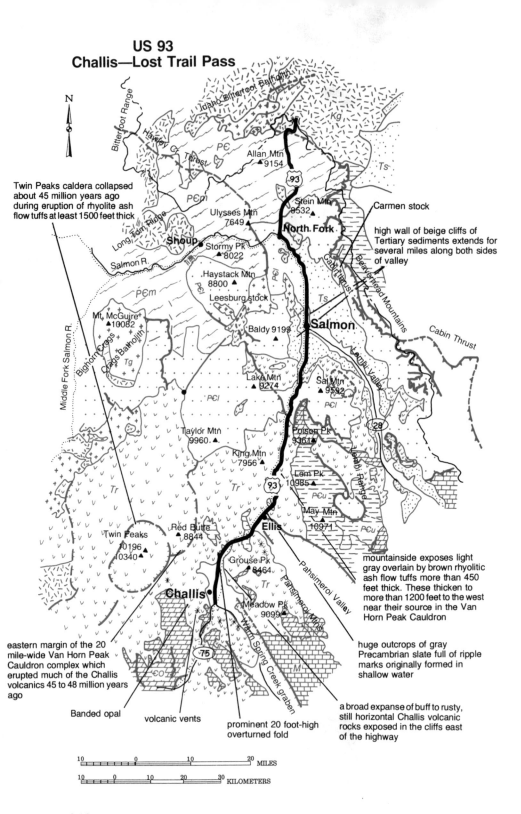

N

Twin Peaks caldera collapsed about 45 million years ago during eruption of rhyolite ash flow tuffs at least 1500 feet thick

Carmen stock

high wall of beige cliffs of Tertiary sediments extends for several miles along both sides of valley

Bitterroot Range

Idaho Bitterroot Batholith

Kg

Hawley Cr.

PЄ

Ts

Allan Mtn
▲9154

93

Stein Mtn
9532▲

North Fork

Long Tom Ridge

PЄm

Ulysses Mtn
7649▲

Shoup

Stormy Pk
▲8022

Salmon R.

PЄm

.Haystack Mtn
8800 ▲

Leesburg stock

PЄl

Ts

Salmon

Cabin Thrust

Beaverhead Mountains

Cabin Thrust

Middle Fork Salmon R.

Mt. McGuire
▲10082

Bighorn Crags

Crags Batholith

Tg

Baldy 9199

Lemhi Valley

28

Laka Mtn
9274

Sal Mtn
▲9592

PЄl

Taylor Mtn
9960.▲

PЄl

Poison Pk
9361▲

Lemhi Range

King Mtn
7956▲

Lem Pk
10985▲

Tr

Tr

93

PЄu

May Mtn
10971

PЄu

Twin Peaks
10196
10340▲

Red Butte
▲8844

Ellis

Grouse Pk
▲8464

Pahsimeroi Valley

mountainside exposes light gray overlain by brown rhyolitic ash flow tuffs more than 450 feet thick. These thicken to more than 1200 feet to the west near their source in the Van Horn Peak Cauldron

Challis

Tr

Pahsimeroi River

Meadow Pk
9099▲

Warm Spring Creek graben

75

M

eastern margin of the 20 mile-wide Van Horn Peak Cauldron complex which erupted much of the Challis volcanics 45 to 48 million years ago

Banded opal

volcanic vents

prominent 20 foot-high overturned fold

huge outcrops of gray Precambrian slate full of ripple marks originally formed in shallow water

a broad expanse of buff to rusty, still horizontal Challis volcanic rocks exposed in the cliffs east of the highway

10   0   10   20 MILES

10   0   10   20   30 KILOMETERS

148

Cliffs eroded in Challis volcanic rocks north of Salmon.

U.S. 93
Challis — Lost Trail Pass
106 miles

Highway 93 follows the Salmon River between Challis and North Fork, the North Fork of the Salmon between North Fork and its headwaters at Lost Trail Pass. Rocks along the way are varied, complicated. Except for a few exposures of hard Precambrian sedimentary rock near the mouth of the Pahsimeroi River around Ellis, all the solid bedrock near the road between Challis and the area about 20 miles south of Salmon is volcanic rock, rhyolite.

Rocks exposed along 20 or so miles of road south of Salmon, and west of the road for about the same distance north of town are Precambrian sedimentary formations. They are hard rocks that the river erodes only with great difficulty to make narrow canyons full of bold outcrops. Watch for the tilted layers of dark mudstone and quartzite. These rocks accumulated as sediments sometime more than one billion years ago, then lay generally undisturbed until the Rocky Mountains formed during late Cretaceous time. That was what tilted the layers.

Precambrian sedimentary rocks around North Fork belong to the Yellowjacket group of formations. Although those rocks are widespread through large parts of central Idaho, this is one of the few places where they appear beside a major highway. The question of how to interpret the Yellowjacket formations has aroused quite a bit of controversy. Many geologists now believe that they are actually Belt rocks. If so, they probably belong at the very base of the sequence, the oldest of the Belt formations. Unlike any other Belt rocks, the Yellowjacket formations contain numerous basalt lava flows.

Except for the sedimentary rocks near North Fork, most of the rocks between North Fork and Lost Trail Pass are basement rocks, ancient gneisses much older than the Precambrian sedimentary rocks near Salmon. At Lost Trail Pass, the road crosses onto the southern edge of a large body of genuine granite, the Joseph pluton.

Rocky Mountain Thrust Faults

Most of the Precambrian sedimentary formations and basement rocks are on one or the other of two big slabs. They almost certainly moved sometime during Cretaceous time, certainly in a generally easterly direction, probably for a distance of some tens of miles, at least. The difficult problems of discovering precisely when, in what direction, and how far those slabs moved all remain to be resolved.

The Cabin thrust fault is the older and lower of the two. It is exposed in the Beaverhead Mountains on the skyline northeast of Salmon, where it places old Precambrian sedimentary formations on top of younger rocks. The Hawley Creek thrust fault, which the road crosses about 15 miles south of Salmon, brought another thick slice of Precambrian rocks onto those above the Cabin fault.

Precambrian Basement Rocks

Watch between North Fork and Lost Trail Pass for occasional road cuts in pale gray rock that has a generally streaky look. It is basement rock. You can get a better view of those rocks along the road that follows the Salmon River west from North Fork. It crosses the Hawley Creek thrust fault onto more Precambrian basement rocks, part of the slab that moved east on the fault. This is the only easy glimpse of the basement rocks of the Salmon River arch.

A number of age dates suggest that these rocks are about 1500 million years old, much younger than other basement rocks in Idaho and nearby southwestern Montana. In fact, that age is only slightly older than some of the Belt sedimentary formations that appear to lie

These white orbicules in the granite near Shoup
are about the size of golf balls.

on top of these basement rocks, actually younger than some of the dates on lava flows in the Yellowjacket formations. It is hard to imagine how those basement rocks could be that young. Perhaps the 1500 million year age is wrong; possibly it marks the time at which the rocks cooled enough to start their radioactive clocks.

Large road cuts just east of Shoup expose an extremely rare and bizarre kind of rock called orbicular granite, a real curiosity. We have seen more spectacular exposures of orbicular granite in the Buffalo Hump area south of Elk City, but these near Shoup are the only ones we know of beside a paved highway. Look for orbicules in the road cuts and in the piles of broken rock between the road and the river. They are approximately spherical masses about the size of golf balls or tennis balls composed mostly of white feldspar. If you examine them closely, you see that the feldspar crystals radiate from the center, and that grains of black biotite define a series of concentric shells within each sphere. Although several theories have been proposed, the origin of orbicules remains unclear, to put it mildly. They seem to form near the edges of granite intrusions. We think they may develop around little fragments of the neighboring rock that were incorporated in the magma, but can't explain how or why that should happen.

The Joseph Pluton

Granite near Lost Trail Pass, the Joseph pluton, appears to lie at the base of the Sapphire block. That is an enormous slice of the earth's upper crust about ten miles thick that moved out of Idaho and about 50 miles into western Montana during late Cretaceous or Eocene time. That movement stripped the top off a large part of the northern part of the Idaho batholith, exposing the newly crystallized granite. This granite at the base of the Sapphire block, well east of the Idaho batholith, suggests a number of interesting ideas.

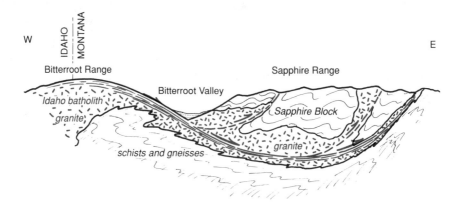

Schematic section showing the Sapphire block moving east into Montana on a sole of molten granite magma.

Ever since geologists discovered overthrust faults, they have wondered how large masses of rock can possibly move on nearly horizontal surfaces. Why doesn't friction prevent the movement? Slabs the size of the Sapphire block, which is about ten miles thick, should create more than enough friction to keep the rocks from moving. If the granite magma of the Joseph pluton was molten when the rocks moved, it could have lubricated the fault surface; in effect, slipped a banana peel under the Sapphire block. If so, all the movement happened while the granite was still molten; it shows very little evidence of shearing. Perhaps crystallization of the granite in the sole of the fault ruined the banana peel effect, thus stopping the eastward movement.

Big crystals of plagioclase feldspar in the Challis volcanic rocks a few miles west of Challis.

The Challis Volcanic Pile

Long stretches of the road pass cliffs and road cuts in pale, rather nondescript Challis volcanic rocks, mostly rhyolite. Watch for exposures of rather massive rock that comes in various pastel shades of gray, yellow, pink, and even green. The rock is layered, but many of the deposits are so thick that you may find the layering hard to see.

Most of that rhyolite is volcanic ash. Some of the less coherent layers, the ones you could almost dig with a shovel, are air fall ash that settled like gently falling snow. A great deal more of that stuff arrived more ferociously in searing clouds of red hot steam laden with ash so hot that it fused together as it settled into more or less solid rock, a welded ash flow. Truly, those must have been appalling eruptions.

They came from several large resurgent calderas in the hills west of the highway. The Twin Peaks caldera about ten miles west of Challis is nearly circular, about ten miles in diameter. It is full of rhyolite and therefore not visible as a depression in the landscape; the geologic map shows it clearly enough. The much larger Van Horn Peak caldera is also full of rhyolite.

153

The top of the hill is rhyolite of the Challis volcanic pile lying on Precambrian sedimentary rocks in the base of the hill. The Salmon River started flowing on the rhyolite, then cut down through the buried hill of older rock.

Those eruptions of about 50 million years ago spread their blankets of ash across a landscape of rugged hills deeply eroded into all the older rocks, including the granites of the Idaho batholith. In many places, the Challis volcanic rocks fill old valleys. The buried stream gravels are now exposed where modern valleys cut across their ancient predecessors.

Basin and Range

Basin and Range faults that trend northwest began slicing the valley and mountain blocks out of this part of the Rocky Mountains between five and ten million years ago. The highway crosses the northern ends of the Pahsimeroi and Lemhi valleys. Both are Basin and Range fault blocks, as are the Big Lost River, Pahsimeroi, and Lemhi ranges.

Section across the highway just north of Challis. The steep Basin Mountains into blocks.

W

Salmon River Mtns.

Twin Peaks
10340 ft.

Parker Mtn.
9151 ft.

Eocene rhyolite

Paleozoic rocks

Idaho batholith

Eocene granite

Eocene granite

Salmon nestles in a broad Basin and Range valley deeply filled with sediments. Watch for the low bluffs of yellowish sediment, most of it so weakly consolidated that someone wanting a fresh sample would do well to bring a shovel. All the dropped blocks in this part of the Basin and Range contain a fill that consists partly of volcanic rocks, largely of sediments deposited while the climate was too dry to maintain streams.

Cobalt

The Blackbird Mine, at the little town of Cobalt, is one of the few in the country capable of producing cobalt, a metal as essential as it is rare. Cobalt is used to produce chemical catalysts and steel alloys that every industrial country needs. The United States normally imports about 70 percent of its cobalt from Africa, gets most of its domestic supply from mines that produce the metal as a by-product.

Ore was discovered at Cobalt sometime late in the last century. The deposit is very large, but the complex ore contains such a variety of minerals that separating cobalt from the other metals, especially copper, is extremely difficult. That is why the Blackbird Mine has never been able to compete on even terms with foreign sources of cobalt. The mine did produce a small amount of cobalt in 1918, then became more or less dormant until the Second World War revived interest in the deposit. Between 1951 and 1959, the mine produced almost 14 million pounds of cobalt, along with substantial quantities of copper, gold, and nickel. Sporadic activity has continued ever since. Large reserves of developed ore await times when economic conditions again make the Blackbird mine competitive.

and Range faults cut the nearly flat fault slices of the northern Rocky

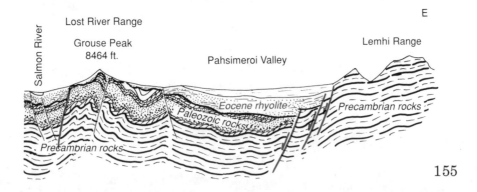

Cliffs eroded in valley-fill sediments southeast of Salmon.

Fluorite

The Meyers Cove area a few miles west of the road to Cobalt is the only significant source of fluorite in Idaho, and it is only barely significant. Mines there produced about 12,000 tons of the mineral during the early 1950s, not much for a non-metallic mineral commodity, then went out of business when the mill burned. The fluorite is in fractures in a granite intrusion that penetrates the Challis volcanic rocks.

Fluorite is typically transparent, crystallizes in nice cubes, and comes in a wide variety of colors that include green, purple, yellow, and pink. It is easy to cleave the crystals into pretty octahedral prisms — watch for them at mineral shows and crafts fairs. Fluorite scratches and breaks much too easily to serve as a gem mineral, but it is useful. Fluorite is the main source of fluorine, has many uses in the chemical industry, and also lowers the melting points of ceramics and smelter slags.

The Yellowjacket Mines

The forlorn town of Yellowjacket lies west of the highway, about 25 winding backwoods road miles from Cobalt. It sprang suddenly into such life as it has had after prospectors found large lode deposits of gold there in 1868. The veins, which also contain silver, copper, and lead, are in Precambrian sedimentary rocks near several igneous intrusions.

Mining began almost immediately after the discovery, but the early milling arrangements were too primitive to support much production.

Installation in 1882 of a large stamp mill driven by a waterwheel greatly increased production; later progressive enlargement of the mill increased it further. Output reached a peak about 1894, then nearly ceased about the turn of the century as the original ore bodies were exhausted. Most estimates place total production during that period at about one million dollars worth of metal, mostly gold. That is such a paltry return for the investment in equipment and labor that it is difficult to imagine how anyone could have made much money in Yellowjacket.

The Yellowjacket district, like most old gold districts, revived for a while with the introduction of cyanide mills about 1910. Then renewed exploration revealed more ore, and Yellowjacket was again a fairly lively place, off and on, until the late 1930s. Nothing has gone very well there since. Increase in the price of gold in the 1980s brought renewed interest in exploration.

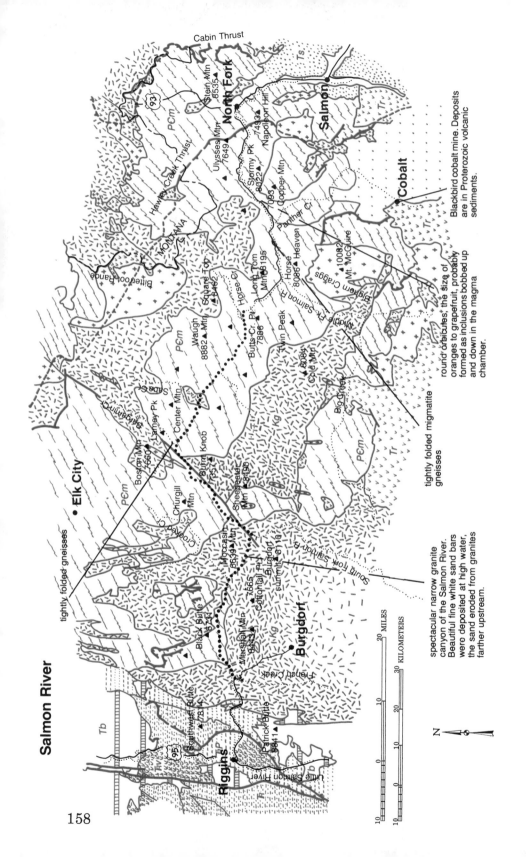

Salmon River

Cabin Thrust

Stein Mtn 8535

North Fork

Ulysses Mtn 7649

Stormy Pk 8022

Napoleon Hill 7493

Salmon

Copper Mtn 7095

Cobalt

Blackbird cobalt mine. Deposits are in Proterozoic volcanic sediments.

Hawley Creek Thrust

MONTANA

Bitterroot Range

PCm

Square Top 8462

Waugh Mtn 8882

Butts Cr. Pk 7886

Long Tom Mtn 8195

Heaven 8086

Horse

Middle Fk Salmon R

Mt. McGuire 10082

Bighorn Craggs

round orbicules, the size of oranges to grapefruit, probably formed as inclusions bobbed up and down in the magma chamber.

Sabe St

Center Mtn

Twin Peak

Cold Mtn 8084

Kg

Big Creek

PCm

Tr

tightly folded migmatite gneisses

Boston Mtn 7660

Lamey Pk

Bluff Knob 7857

Sheepeat Mtn 8490

Churgill Mtn

Crooked C.

Elk City

PCm

tightly folded gneisses

Moccasin 8639

Cottontail Pt 7665

Burgdorf summit 8110

South Fork Salmon R

Black Butte 5243

Marshall Mtn 8243

Friend Creek

Burgdorf

spectacular narrow granite canyon of the Salmon River. Beautiful fine white sand bars were deposited at high water, the sand eroded from granites farther upstream.

Tb

Southwest Butte 7814

Patrick Butte 8841

Riggins

Little Salmon River

N

MILES

KILOMETERS

158

Why it is called "The River of No Return"

THE SALMON RIVER CANYON

We offer this section to our frustrated readers who float the Salmon River Canyon on days when the fish aren't biting. You can watch the rocks, knowing that very few geologists have ever studied them.

The Salmon River Canyon, the "River of No Return," thwarted the first attempt of the Lewis and Clark expedition to find a river route to the Pacific Ocean; they weren't interested in whitewater boating. The forbidding canyon has been an obstacle ever since. Its canyon still severs northern Idaho from the more densely populated southern part of the state. No road approaches the Salmon River Canyon for more than eighty miles of river; no bridge violates its barrier. Only a few people, a mere handful, live anywhere near this gorge that defines and defends so much of the wilderness of central Idaho.

This remote region of Idaho owes its isolation partly to its rugged landscape, partly to its apparent lack of significant mineral resources, as much as anything else to the devastating forest fire of August, 1910 that destroyed its timber resources. The remoteness and inaccesibility of this large region of central Idaho, along with its lack of mineral resources, explains why its rocks have received relatively little attention.

Gneiss exposed beside the Salmon River about seven miles below Shoup.
—U.S. Geological Survey photo by C.P. Ross

The Canyon

As in much of central Idaho, the hills north and south of the Salmon River Canyon are softly rounded, richly upholstered in a thick mantle of soil and deeply weathered rock. Although it is difficult to find any good evidence that might exactly betray the age of those profoundly weathered hills, it is even more difficult to suppose that they could be young.

The Salmon River Canyon seems much younger. It cuts a deep and narrow slot through those softly billowing hills; leaves large expanses of bedrock nakedly exposed in its walls, virtually without a cloak of soil. We strongly suspect that the modern Salmon River, like many other streams in the region, began to flow about 2.5 million years ago through a valley that was first eroded during the wet climatic period of late Miocene time, then abandoned during the long dry spell of Pliocene time. If so, patches of Pliocene gravel should survive here and there on the upper canyon walls as evidence that they were already low and ready to collect sediment while the climate was dry.

The Salmon River Arch

For most of the eastern part of its canyon length, the Salmon River follows the Salmon River arch, an enigmatic expanse of Precambrian basement rock, old continental crust, that separates the northern and southern parts of the Idaho batholith. There is no doubt that the rocks exposed in this long reach of the canyon belong to the Precambrian basement complex, the most ancient part of the continental crust.

160

Most of those rocks are gneisses and schists, metamorphic rocks that formed through recrystallization of much older rocks under very high temperatures, a red heat.

There is some considerable doubt about the age of those ancient basement rocks, and how to interpret them. The best available age dates place the period of their metamorphism at about 1500 million years ago. That is just barely old enough to qualify those rocks as the continental crust on which the Precambrian sedimentary formations of central Idaho accumulated, not nearly old enough to fit them into the regional picture of much older age dates on basement rocks. Those age dates seem too young. Perhaps their radioactive clocks were not started until the basement rocks cooled enough or the clocks were partially reset by heat from the Idaho batholith.

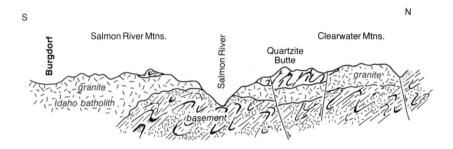

Section across the western part of the Salmon River Canyon about seven miles west of Crooked Creek.

Some geologists, including ourselves, regard the Idaho batholith as a sheet of granite several thousand feet thick, an intrusion of magma that rose to a crustal level where it found native rocks with the same density as the magma, then spread horizontally. Those geologists explain the Salmon River arch as a broad anticline that warped the granite sheet of the Idaho batholith upward, thus exposing it to erosion. They regard the metamorphic rocks of the Salmon River arch as older basement rocks beneath the batholith.

Metamorphic rocks in the Salmon River Canyon.
— U.S. Geological Survey photo by F.W. Cater

The Metamorphic Rocks

Watch for the exposures of metamorphic rocks along the eastern half of the canyon, everywhere east of the mouth of Bargamin Creek and most of the way between there and the mouth of the South Fork of the Salmon River. Most of the basement rocks are gneisses that suggest streaky looking granites and schists that contain enough mica to make them flaky. Dark banding in the dominantly pink and gray gneisses swirls through the outcrops. Freshly broken surfaces on the darker schists glitter with flakes of mica, mostly the black variety called biotite.

Section across the Salmon River Canyon

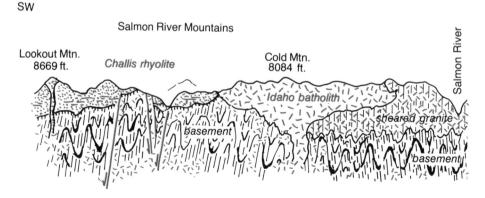

162

Close examination of the gneisses and schists will show that their mineral grains tend to lie parallel to each other. Most of the mica flakes, for example, lie with their flat surfaces facing in the same direction, as though they were sheets of paper lying on the floor. With a few conspicuous exceptions, such as the bright red crystals of garnet, the other minerals in metamorphic rocks show the same tendency. That, more than anything else, is what gives so many metamorphic rocks their streaky appearance and flaky quality.

The Idaho Batholith

Except for the area of metamorphic rocks between Crooked Creek and French Creek, most of the canyon west of the South Fork is eroded in granite of the Atlanta batholith, the southern part of the Idaho batholith. Those rocks are about 75 million years old. Watch for exposures of massive pale gray granite — among geologists, massive means that a rock lacks a preferred orientation of mineral grains. Granites lack the streaky look characteristic of most metamorphic rocks; turn a piece of granite over, and the rock looks the same viewed from the new direction.

Like granites everywhere, those in the Idaho batholith consist essentially of feldspar crystals intergrown with much smaller numbers of quartz grains. Scattered grains of a black mineral, typically either flakes of biotite mica or needles of hornblende, pepper most granites. A few contain flakes of white mica. Most of the feldspar in the Idaho batholith forms somewhat blocky grains that are milky white or tinged with pink. Quartz grains are glassy, but often look dark in the rock because you see right through the clear mineral into its shadowed setting in the rock. Distinguish between biotite and hornblende by checking to see whether the black grains are flat flakes or stubby needles.

about seven miles west of Corn Creek.

163

The Continental Suture

A few miles west of the mouth of French Creek, the river cuts its canyon through the suture zone between the former western margin of the North American continent and the old chain of volcanic islands that joined the continent to become the Seven Devils complex.

Watch for a change from the pale gray granite of the Idaho batholith to much darker rocks that still consist mostly of feldspar, but contain little or no quartz and much more black minerals. Also watch for a change from massive rocks to mylonite in this case granitic rocks that have been sheared into a layered gneiss. That shearing happened as the oceanic crust slid beneath the old continental margin.

The Seven Devils Complex

In the last 20 miles or so above Riggins, the Salmon River Canyon cuts through rocks of the Riggins group, probably the metamorphosed equivalent of the Seven Devils complex, the old chain of volcanic islands that formed somewhere out in the Pacific Ocean, then crashed into North America. These are mostly volcanic rocks, mostly gray or

A schematic section across Idaho of 100 million years ago to show the sinking oceanic crust, the western Idaho mylonite zone and the future site of the Idaho batholith.

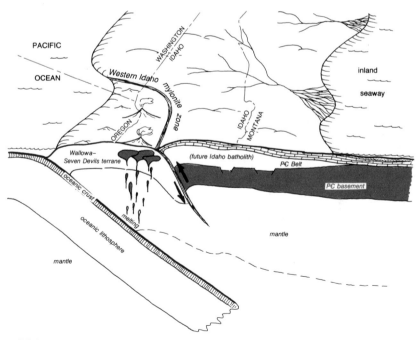

greenish gray, but the complex also contains sedimentary rocks, sandstones, shales, and a bit of limestone. All are now metamorphosed into slates and schists. The metamorphism makes it difficult be sure just what the original rock may have been.

Watch the platy layering in the sheared granitic rocks and the layering in the metamorphic rocks of the Riggins group to see that they tilt down to the east. The sinking oceanic crust apparently dragged them into that attitude as it added the Seven Devils complex to the western margin of North America.

Age dates on the volcanic rocks and fossils in the sedimentary rocks show that the volcanic islands that became the Seven Devils complex formed during Permian and Triassic time, roughly between 220 and 260 million years ago. Age dates on the dark igneous rocks in the suture zone between the old islands and North America show that they formed about 100 million years ago. Those are the rocks that welded the Seven Devils complex to North America, so their age tells us when the two finally joined.

US 95
Payette—New Meadows

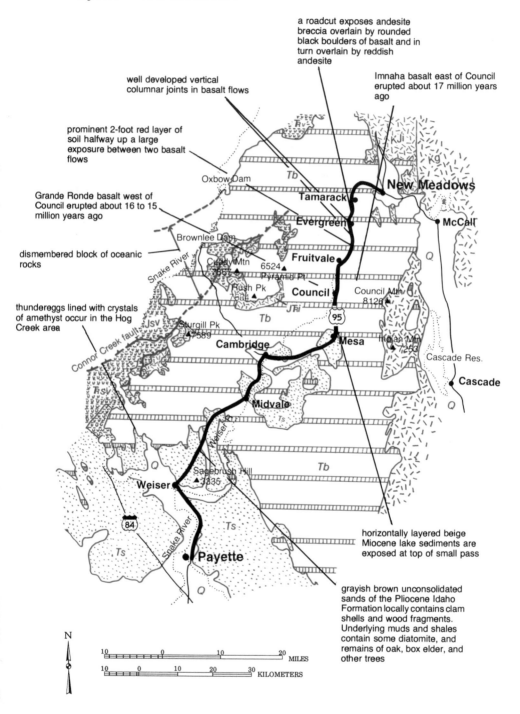

a roadcut exposes andesite breccia overlain by rounded black boulders of basalt and in turn overlain by reddish andesite

Imnaha basalt east of Council erupted about 17 million years ago

well developed vertical columnar joints in basalt flows

prominent 2-foot red layer of soil halfway up a large exposure between two basalt flows

Oxbow Dam

Tb

New Meadows

Tamarack

Grande Ronde basalt west of Council erupted about 16 to 15 million years ago

Brownlee Dam

Evergreen

McCall

dismembered block of oceanic rocks

Cuddy Mtn
7867

6524

Pyramid Pt

Fruitvale

Q

Snake River

Rush Pk
7588

Council

Council Mtn
8126

thundereggs lined with crystals of amethyst occur in the Hog Creek area

Jsv

JRi

Tb

Mesa

Indian Mtn
7253

Sturgill Pk
7589

Cambridge

Cascade Res.

Connor Creek fault

Rsv

Cascade

Q

Midvale

Ts

Weiser River

Q

Tb

Sagebrush Hill
3335

Weiser

Ts

84

Snake River

Ts

Q

Payette

horizontally layered beige Miocene lake sediments are exposed at top of small pass

grayish brown unconsolidated sands of the Pliocene Idaho Formation locally contains clam shells and wood fragments. Underlying muds and shales contain some diatomite, and remains of oak, box elder, and other trees

Rv

KJi

Kg

95

N

10	0	10	20

MILES

10	0	10	20	30

KILOMETERS

U.S. 95
Payette — New Meadows
88 miles

All rocks exposed along the road between Payette and New Meadows are either basalt lava flows of the Columbia Plateau or sedimentary rocks closely associated with them. All are about 15 million years old. It is easy to tell them apart: the basalt is very hard and nearly black; the soft sedimentary rocks come in a variety of pale yellowish colors.

The forested West Mountains that rise in the distance ten to 15 miles east of the highway are eroded into much older granites of the Idaho batholith. Those mountains are in the old North American continent. Mountains west of the road are eroded in the exotic rocks of the Seven Devils complex, the islands that came in from the ocean about 100 million years ago. The old continental margin must lie concealed somewhere beneath the flood basalt flows.

Columbia Plateau basalt flooded the valleys of the mountains east and west of the highway. Those lava floored valleys show that the landscape of western Idaho must look about as it did before the Columbia Plateau erupted, except for all the basalt.

Plateau Basalt

Except for a short stretch near Weiser, the entire route crosses the Weiser embayment, the southernmost of three large lobes of the Columbia Plateau that extend east into Idaho. Those three areas were too low to stand above the floods of lava.

Stacked flows of basalt exposed in the canyon of the Weiser River about 4 miles south of Cambridge. — U.S. Geological Survey photo by C.P. Ross

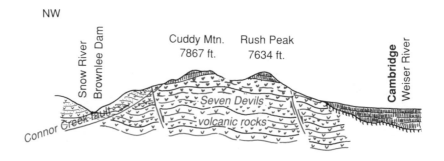

NW

Section across the line of

The first lava flows to flood the Weiser embayment were the Imnaha basalts, which erupted between 16 and 17 million years ago. We interpret those early floods of Imnaha basalt as the first overflows from the lava lake that filled the meteorite explosion crater in southeastern Oregon after it formed about 17 million years ago. At least 19 flows of Imnaha basalt flooded some parts of the Weiser embayment, fewer in other parts. They filled the old stream valleys, and in some places piled up to a thickness of more than 3000 feet. The best exposures are in places where the modern streams carved deep valleys into them.

People with a keen eye for the subtleties of basalt recognize the Imnaha flows by looking in the fresh rock for tiny grains of glassy green olivine and much larger flat crystals of white plagioclase feldspar. Weathering quickly converts the olivine into little dabs of rusty clay. You can almost recognize Imnaha basalts from a distance because they tend to weather rather easily into rounded boulders, and long, relatively smooth slopes.

Next came the Grande Ronde basalts, which covered much of the Imnaha basalt. Most of the Grande Ronde flows erupted between 15 and 16 million years ago. These are the flows you see across most of

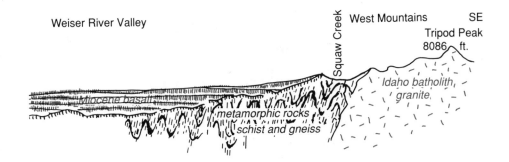

Weiser River Valley

Squaw Creek

West Mountains · · · SE

Tripod Peak
8086 ft.

Idaho batholith, granite

Miocene basalt

metamorphic rocks
schist and gneiss

the highway at Cambridge.

the northern Columbia Plateau in its enormous spread across northern Oregon and eastern Washington.

The Grande Ronde flows erupted from long fissures in the area just west of the Weiser embayment. We think the molten magma flowed north through the fissures from the lava lake that filled the crater in southeastern Oregon. Now, those fissures are large basalt dikes, thousands of them, that trend generally north. They are the spectacular Joseph dike swarm, the plumbing of the northern Columbia Plateau.

Although westernmost Idaho contains plenty of those big dikes, it is hard to find easily accessible places to see them. The best place is along Oregon Highway 3, where it passes through the Grande Ronde Canyon just south of the Washington line. Dozens of basalt dikes make dark streaks on the canyon walls and stand on the lower slopes in long ridges that look like ruined walls.

Grande Ronde basalt contains more silicon and less aluminum than the Imnaha basalt. You see no green crystals of olivine in the fresh rock, no big flat grains of white plagioclase feldspar. Grande Ronde basalt is distinctive even from a distance because it resists

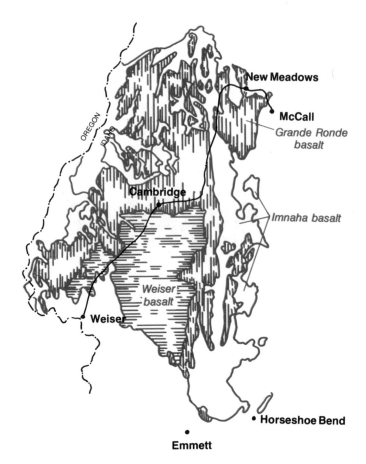

The distribution of different basalt types in the Weiser embayment.

weathering well enough to make prominent ledges of crisp vertical columns of basalt in the hillsides, with talus slopes of angular debris beneath them. Soils developed on Grande Ronde basalt tend to be reddish.

The Weiser basalts are the youngest volcanic rocks in the embayment, and they have little in common with the older flood basalts. Most of those late flows erupted locally from two small groups of volcanoes: One is about 12 miles west of Cambridge, the other overlooks the Weiser River a few miles northeast of Cambridge. Weiser basalt flows were considerably less voluminous than those that came before, nothing like the overwhelming floods of Imnaha and Grande Ronde basalt. And the Weiser basalt lavas must have been fairly viscous because the flows did not spread very far. Many formed ash, or broke up into rubbly masses of basalt, instead of spreading out into smooth lava flows.

Mercury

The Idaho-Almaden Mine, an open pit on top of Nutmeg Mountain about 11 miles east of Weiser, produced more than half of Idaho's total output of mercury in just six years.

The ore is in sandstone deposited in Lake Payette, one of the largest of those lakes impounded behind basalt lava flows during late Miocene time, when the Columbia Plateau was active. Sometime during the last two million years, circulating hot water, probably a hot spring, altered much of the sandstone on Nutmeg Mountain to common opal heavily laced with cinnabar, a mercury mineral.

Opal is a formless and non-crystalline variety of silica closely related to quartz, a fairly common mineral in and around many kinds of volcanic and altered rocks. It normally forms white or yellowish green masses that rarely attract a second glance. Tradition has it that a sheepherder noticed outcrops of the Nutmeg Mountain ore in 1936. If so, he must have been remarkably observant, or maybe just lucky.

Cinnabar, the mercury sulfide mineral, is bright red in freshly broken rock, and obvious enough. Long exposure to the sun darkens it into gray stuff that rarely attracts much attention. Weathered outcrops full of cinnabar are so easy to overlook that mercury deposits are generally the last ores prospectors find. Whoever found the mercury deposit on Nutmeg Mountain must have been breaking the rocks to discover the bright pink ore within.

Cinnabar is so unstable that strong heating breaks it down into shiny little drops of liquid metal. Mercury mines roast the metal out of the ore, then condense it in very pure form from the escaping vapor. Mercury production at Nutmeg Mountain began in 1939, then ceased in 1942, after the mine had produced some 4000 flasks of mercury, each a steel container filled with 76 pounds of the shiny liquid metal. The property went back into operation with new refining equipment in 1955 and produced another 12,000 flasks of mercury by 1961, when the ore finally ran out. In its better years, the mine roasted about 25 tons of ore to fill each flask.

Gypsum

The hills about 30 miles northwest of Weiser contain deposits of gypsum, a calcium sulfate mineral valuable mainly as a soil conditioner and as raw material for sheet rock and plaster of paris. Mines working there have produced small amounts of gypsum in Idaho, much more on the Oregon side of the Snake River.

The gypsum is in layers within shale that was deposited during Triassic time, some 180 or so millions of years ago. The shales are part of the Seven Devils complex, so those gypsum beds were laid down on volcanic islands somewhere out in the Pacific Ocean, then added to Idaho about 100 million years ago — an unconventional way to import mineral commodities.

This volcanic mudflow in the Salmon River Canyon was once part of an island far out in the Pacific Ocean. —U.S. Geological Survey photo by W.B. Hamilton

U.S. 95
New Meadows — Lewiston
160 miles

This route traces a line close to the former western margin of North America, wavering between rocks that belong to old North America and others that are parts of islands that came in from the Pacific Ocean. Along much of the way, the highway crosses basalt lava flows erupted during late Miocene time, long after those islands joined North America. The details of which rocks appear where along this road are fairly complex.

A Geologic Itinerary

The route between New Meadows and Pinehurst follows the Little Salmon River near the ragged eastern edge of the Columbia Plateau. Both communities stand on basalt, as does much of the area between them. The fairly flat landscape with its occasional dark outcrops of black basalt tell you that this is indeed the Columbia Plateau.

About midway between New Meadows and Pinehurst, the road skirts the western edge of a large mass of granite, the top of a mountain that stands half buried in basalt. North of the granite, the

173

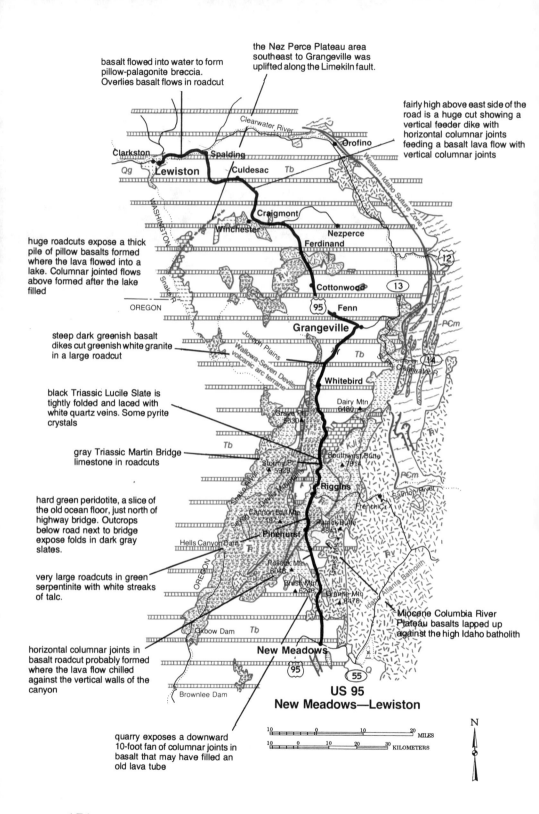

the Nez Perce Plateau area
southeast to Grangeville was
uplifted along the Limekiln fault.

basalt flowed into water to form
pillow-palagonite breccia.
Overlies basalt flows in roadcut

fairly high above east side of the
road is a huge cut showing a
vertical feeder dike with
horizontal columnar joints
feeding a basalt lava flow with
vertical columnar joints

Clearwater River

Clarkston
Spalding
Orofino
Lewiston
Culdesac
Tb
Qg
WASHINGTON

Craigmont
Winchester
Nezperce
Ferdinand
12

huge roadcuts expose a thick
pile of pillow basalts formed
where the lava flowed into a
lake. Columnar jointed flows
above formed after the lake
filled

Snake R.
Tv
Cottonwood
13
95 Fenn

OREGON
PCm

Grangeville

steep dark greenish basalt
dikes cut greenish white granite
in a large roadcut

Joseph Plains
Wallowa-Seven Devils
volcanic arc terrane
Tb
14
Clearwater R.

Whitebird

black Triassic Lucile Slate is
tightly folded and laced with
white quartz veins. Some pyrite
crystals

Dairy Mtn
6480

Grave Pt.
5630
Tb

gray Triassic Martin Bridge
limestone in roadcuts

Southwest Butte
7814
PCm

Stormy Pt.
5929
Riggins

hard green peridotite, a slice of
the old ocean floor, just north of
highway bridge. Outcrops
below road next to bridge
expose folds in dark gray
slates.

French Cr.
Salmon River
Cannon Ball Mtn
7197
Patrick Butte
8817

Hells Canyon Dam
Pinehurst
Tv

very large roadcuts in green
serpentinite with white streaks
of talc.

Rollock Mtn
8046
Granite Mtn
8478

Miocene Columbia River
Plateau basalts lapped up
against the high Idaho batholith

Idaho Atlanta Batholith
Brush Mtn
6228

horizontal columnar joints in
basalt roadcut probably formed
where the lava flow chilled
against the vertical walls of the
canyon

Oxbow Dam
Tb

New Meadows

Brownlee Dam
95
55
Q

**US 95
New Meadows—Lewiston**

quarry exposes a downward
10-foot fan of columnar joints in
basalt that may have filled an
old lava tube

10   0   20   MILES
10   0   10   20   30   KILOMETERS

N

Basalt palisades in the walls of the Snake River Canyon about fifteen miles above Lewiston. —U.S. Geological Survey photo by I.C. Russell

road crosses small areas of metamorphic rock, as well as quite a lot of Columbia Plateau basalt. The metamorphic rocks belong to the islands that joined North America about 100 million years ago. Between Pinehurst and Riggins, the road follows the Little Salmon River to its mouth in the Salmon River across a wild assortment of those exotic island rocks: the Seven Devils complex and Riggins group. Essentially the same rocks appear in the depths of the Salmon River Canyon between Riggins and Whitebird.

Along most of the route between Whitebird and Lewiston, the highway crosses basalt lava flows of the Columbia Plateau. The extraordinarily fertile wind blown silt of the Palouse Hills covers most of the basalt, but the distinctive palisades of brownish columns appear here and there in small valleys crossing the road. Canyon walls above the Clearwater River east of Lewiston provide almost continuous exposures of the basalt. In the midst of all that basalt, the road passes through the top of another granite hill almost buried in the floods of basalt. Look for the granite in a small area about midway between Fenn and Craigmont.

The Riggins Group and the Seven Devils Complex

The spectacular drive through the depths of the Salmon River Canyon between Riggins and Whitebird passes equally outstanding rocks, the squashed remains of the group of oceanic islands that joined

175

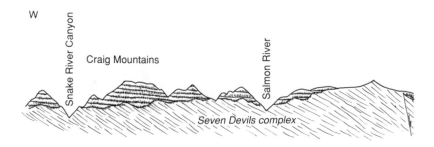

W

Snake River Canyon

Craig Mountains

Salmon River

Seven Devils complex

Section across the highway a

North America sometime around 90 to 100 million years ago. The canyon opens a window through the Columbia Plateau into the rocks that probably lie hidden beneath the basalt throughout much of western Idaho.

These nearly obliterated oceanic islands were part of a large group that formed somewhere in the Pacific about 200 million years ago, during Permian and Triassic time. They joined North America during middle Cretaceous time, about 100 million years ago. That was when they were squashed — notice that the layering stands at steep angles all through the canyon. Scattered exposures of the same rocks appear along the Little Salmon River between Riggins and New Meadows.

In general, the more thoroughly metamorphosed rocks in this westernmost part of Idaho, the gneisses and schists along the northern part of the route, belong to the Riggins group. The less recrystallized rocks that still retain something of the appearance of the original volcanic and sedimentary rocks are part of the Seven Devils complex. The exact relationship between them is not clear, except that they appear to be closely associated.

Despite those uncertainties, it does seem clear enough that the rocks of the Seven Devils complex were dragged under the Riggins group along faults that dip down to the east. That happened as they rode the sinking slab of lithosphere into the trench that existed off the west coast of Idaho until about 100 million years ago.

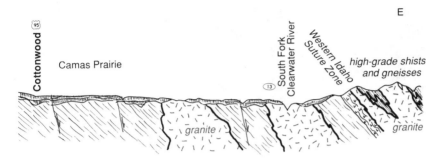

few miles north of Grangeville.

The best exposures of Riggins group rocks are in the area a few miles north and south of Riggins. Watch for slabby road cuts in gray schist, rocks thoroughly enough metamorphosed that it is difficult to tell what they were before they were baked, squashed, and recrystallized.

Serpentinite

You can see nice exposures of greenish black peridotite in a road cut just north of the Goff Bridge, about two miles north of Riggins. And don't miss the big road cuts in serpentinite in the area between five and eight miles south of Riggins. These are among the few places in Idaho where you can see either rock.

Peridotite is a dense rock composed mostly of black augite and little green crystals of olivine, which tend to break down into reddish bits of clay. It is the most abundant rock in the Earth, almost certainly the kind of rock that forms most of the mantle. If peridotite reacts with water, it turns into serpentinite.

Serpentinite is a dense and darkly greenish rock that tends to form rather blocky exposures. Many of the blocks have polished surfaces covered with a shiny coating of green and yellowish green serpentine minerals. Most serpentinite forms deep beneath oceanic ridges where sea water sinks into fissures, then reacts with very hot peridotite in

Platy schist, part of the former volcanic islands in the Salmon River Canyon. —U.S.Geological Survey photo by W.B. Hamilton

the upper part of the Earth's mantle. Like all the rest of the sea floor, that stuff eventually finds its way to an oceanic trench.

Serpentinite is a very weak rock, with a tendency to be slippery. Large masses of it squeeze out of the slab sinking into the trench and work their way up along fractures in the rocks accumulating in a trench. Geologists often refer to that tendency of serpentinite to slither through faults and fractures as "watermelon seed tectonics." That is how this serpentinite found its way into the rocks of the Seven Devils complex and Riggins group.

White streaks in the dark green serpentinite are talc, the mineral source of talcum powder. Good quality talc is quite valuable, but this stuff is less than worthless because it contains asbestos, which causes cancer about as certainly as anything. Don't try making your own talcum powder in the Salmon River Canyon. Most commercial talc now comes from southwestern Montana, where the deposits contain no asbestos.

The black Lucile slate appears in several road cuts around Lucile —where else? The rock is black because it contains so much graphite, the mineral used in pencil leads, that it easily marks a sheet of paper. Some exposures of the Lucile slate also contain nicely formed cubes of brassy yellow pyrite, fool's gold. Like the other rocks exposed in this part of the Salmon River Canyon, the Lucile slate was tightly folded as these old islands mashed into the old oceanic trench along the former western edge of North America.

Pale gray limestone exposed in large cuts just south of Lucile is the Martin Bridge formation. It seems to have formed in reefs that surrounded some of the oceanic islands. In places where the Martin Bridge limestone is less thoroughly deformed than it is in this area, it contains abundant fossils of animals that lived during Triassic time, about 220 million years ago. But those fossils are quite unlike the remains of Triassic animals in limestones on the North American continent, more like those in southeast Asia. The exotic fossils in the Martin Bridge limestone are further evidence that these rocks in the Salmon River Canyon formed far away from where we now see them.

Detailed studies of the fossils in the Martin Bridge limestone show that they include the remains of a large number of different species, which vary from place to place. That suggests that the formation was laid down around a number of separate islands, each with its own distinctive assortment of animals. Reefs that fringe continents tend to contain the same animals from one end to the other.

Columbia Plateau Basalt

Many of the biggest lava flows in the Columbia Plateau erupted just west of the area where the boundaries of Idaho, Oregon, and Washington meet. The deep river canyons there cut through thousands of basalt dikes that trend generally north. Those dikes are the filled fractures through which the basalt magma erupted, the plugged plumbing, as it were, of the Columbia Plateau. A few of those big dikes appear in the valleys of westernmost Idaho. The easiest way to spot

This gorgeous palisade of basalt columns exposed on Race Creek just northwest of Riggins is the lower part of a lava flow. The rubbly material above is the irregularly fractured middle of the flow. If the cut were a bit higher, it would expose a second row of columns at the top of the flow.

Cordwood jointing in a large basalt dike on Squaw Creek southwest of Riggins. The joints formed as the crystallizing rock fractured perpendicular to the contacts.

them is to watch for exposures of basalt in which the columns lie horizontally like stacked cordwood, instead of standing vertically as they typically do in lava flows.

Exposures in the deep canyons also show that the earlier flows backed up into the mountain valleys of western Idaho, filling some of them with hundreds of feet of basalt. Only after those early flows had fairly well leveled the landscape could the later flows spread out into thin sheets of basalt that cover large areas. Now, some of the modern

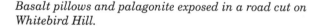

Basalt pillows and palagonite exposed in a road cut on Whitebird Hill.

canyons cut across those older valleys, exposing their very deep fills of basalt. Unfortunately, none of those are easily visible from this highway. Watch for them if you drive the side roads, and along Highway 12 near Kooskia.

While the Columbia Plateau was building, the streams draining central Idaho flowed generally west, as they still do. Wherever one of the enormous floods of basalt crossed a streams, it impounded another shallow lake, which laid down layers of sediment that now appear as white layers sandwiched between flows, the Latah formation.

Most of those lakes ended violently as a later lava flow filled them in a holocaust of steam and ash. Molten basalt erupts at a temperature of about 2000 degrees fahrenheit, a bright red heat. When a basalt flow pours into a lake, the water explodes into clouds of steam that blast much of the lava into shreds of sizzling ash. Reaction between the hot ash and steam converts much of the basalt into a nondescript mass of yellowish and brownish mud called palagonite. It typically contains embedded chunks of fresh black basalt about the size and shape of pillows.

Lava flows exposed in the wall of the Salmon River Canyon. —U.S.Geological Survey photo by I.C. Russell

Old river terraces stand high above the Little Salmon River, a few miles south of Riggins. The river kept pace with uplift, leaving its old floodplain standing high as terraces.

The Seven Devils Mountains

The Seven Devils Mountains are a narrow ridge that separates the parallel deep canyons of the Salmon and Snake rivers. The rocks near the river belong to the Seven Devils complex. Parts of the ridge crest are basalt lava flows, the highest part of the Columbia Plateau.

It is difficult to imagine that the high Seven Devils Mountains could have existed when the rivers on either side began flowing. The route those two streams now follow must then have been the lowest available path. We suspect that the two deep canyons follow the paths of spillways that drained large lakes impounded behind lava flows during late Miocene time.

Why are the Seven Devils Mountains so high? They certainly owe part of their height to Basin and Range faulting. We suspect they owe another part of their high elevation to the deep canyons on either side. To some extent, erosion of the canyons helped raise the mountains. Erosion of Hells Canyon certainly raised the canyon rim on the Oregon side.

The earth's crust floats on the red hot rocks of the mantle beneath just as an air mattress floats on water. A heavy load, such as a large glacier or volcano, placed on the crust sinks it deeper into the viscous

182

rocks of the mantle. If a load is removed from the crust, it rises as the mantle rocks slowly flow in beneath to replace the material removed from the surface.

As the Snake and Salmon rivers eroded their deep canyons, the earth's crust rose, floated, carrying the narrow ridge of the Seven Devils Mountains higher. That is possible because erosion has been removing material from the canyons, not from the surface of the Columbia Plateau. This is one area where stream erosion is causing the difference in elevation between mountain top and canyon floor to increase and the tops of the mountains to rise. Erosion doesn't necessarily wear the mountains down, as we tend to assume. If it concentrates on the valleys and spares the ridge crests, erosion may well have precisely the opposite effect.

We think you can see the interaction between erosion and uplift along Highway 95 immediately north of Whitebird Hill. The plateau surface there quite visibly tilts up to form a low welt around the canyon rim. Furthermore, the small valleys that corrugate the broad surface east of the road down Whitebird Hill are steepest on their downslope sides, precisely what you would expect if the entire surface has been tilting steeper as they were eroded. We think that tilting reflects the uplift in response to erosion of the canyon.

The curiously unsymmetrical small valleys suggest that this surface has been tilting down to the right since the streams started to erode the little valleys. View east from Whitebird Hill.

Florence

The little that remains of the old town of Florence lies tucked away in the hills north of the Salmon River about 16 miles directly northeast of Riggins — the winding road mileage is much greater. Florence had its day. Prospectors found placer gold there in the late summer of 1861, and the usual horde of otherwise unemployed but incorrigibly hopeful miners followed, more than 1000 of them by November. Most of those left as the winter set in, and most of the few who stayed probably wished they had left. Truly, It was a desperate winter: far too much snow and not nearly enough to eat. A large proportion of the population was down with scurvy by mid-winter.

By the time the snow began to melt after that horrible winter in the mountains, something like 10,000 miners had flocked to Florence. The placer mining was very good during that blossoming summer of 1862, but surely not nearly good enough to provide something for everyone in the immense horde of miners. The majority of those miners worked as more or less miserable wage hands for the lucky minority who had arrived early enough to stake one of the good claims, of which there were not nearly enough to go around.

U.S. 95
Lewiston — Coeur d'Alene
115 miles

Lewiston is on the eastern edge of the Columbia Plateau, and so is Coeur d'Alene. The route between those cities stays close to the boundary between the Columbia Plateau and the Northern Rockies, edging first to one side, then to the other. Most of the rocks right along the highway are Precambrian sedimentary formations intruded in places by granite of the western part of the Idaho batholith. The Columbia Plateau is a constant presence in the distance west of the highway, here and there along the highway. Watch for occasional outcrops and road cuts that expose the black basalt lava flows.

Moscow is on a small eastern embayment of the Columbia Plateau, a place where the basalt flooded eastward into a broad valley. Granite hills that embrace that old valley north, south, and east of Moscow stood above the floods of basalt. The granite a westward outlier of the Idaho batholith. Granite extends to a mile or two north of Viola, where the road crosses onto hills eroded into much older Precambrian sedimentary rocks, now more or less recrystallized into metamorphic rocks. Those ancient rocks make all the hills between Viola and Coeur d'Alene Lake.

Ancient Sedimentary Rocks

Here and there along the road between Potlatch and Coeur d'Alene, you see scattered exposures of Precambrian sedimentary rocks. Although the point has been the subject of some debate, it seems likely that these are Belt formations. Many isolated hills that rise above the broad surface of the Columbia Plateau consist of similar rocks. Sedimentary rock types range from dark gray mudstones to pink sandstones.

Granite and Metamorphic Rocks

Schists and gneisses exposed near the north end of the route could be old basement rock, but it seems more likely that they formed through metamorphism of Belt sedimentary formations. It is also possible that they could be exotic rocks, parts of islands that came in from the Pacific to join North America. In any case, they certainly are thoroughly metamorphosed sedimentary rocks, which look very much alike regardless of their age, or where they came from.

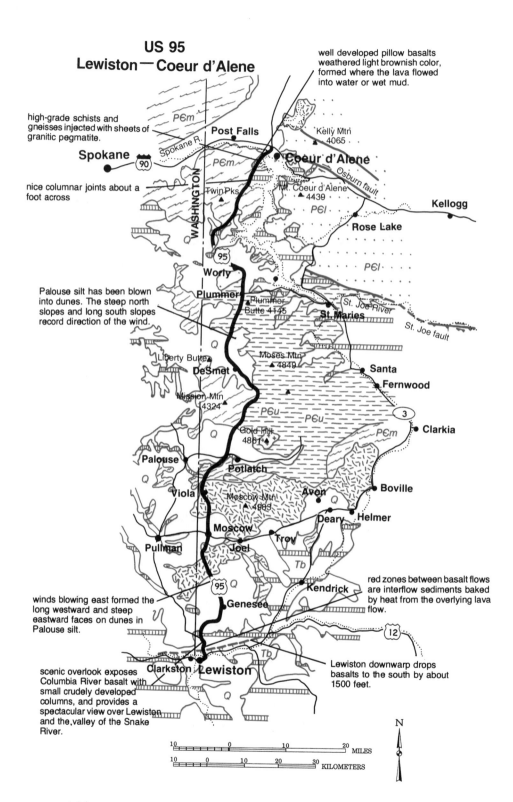

US 95
Lewiston—Coeur d'Alene

well developed pillow basalts weathered light brownish color, formed where the lava flowed into water or wet mud.

high-grade schists and gneisses injected with sheets of granitic pegmatite.

Spokane (90)

nice columnar joints about a foot across

PЄm

Spokane R.

WASHINGTON

Post Falls

Kelly Mtn 4065

Coeur d'Alene

Osburn fault

Twin Pks

Mt. Coeur d'Alene 4439

Kellogg

PЄm

PЄl

Rose Lake

(95)

Worly

PЄl

Plummer

Plummer Butte 4145

St. Joe River

St. Maries

St. Joe fault

Palouse silt has been blown into dunes. The steep north slopes and long south slopes record direction of the wind.

Q

Liberty Butte

Moses Mtn 4849

Santa

DeSmet

Fernwood

Mission Mtn 4324

PЄu

PЄu

(3)

Q

PЄm

Clarkia

Gold Hill 4881

Palouse

Potlatch

Q

Viola

Moscow Mtn 4983

Avon

Boville

Q

Deary

Helmer

Moscow

Pullman

Joel

Troy

Tb

red zones between basalt flows are interflow sediments baked by heat from the overlying lava flow.

(95) Q

Kendrick

winds blowing east formed the long westward and steep eastward faces on dunes in Palouse silt.

Genesee

(12)

Lewiston downwarp drops basalts to the south by about 1500 feet.

Tb

scenic overlook exposes Columbia River basalt with small crudely developed columns, and provides a spectacular view over Lewiston and the valley of the Snake River.

Clarkston **Lewiston**

Q

N

```
10        0        10        20
|=========|========|========| MILES
10   0    10    20    30
|===|====|====|====| KILOMETERS
```

186

Those metamorphic rocks were reheated and to some extent recrystallized during Cretaceous time, while an oceanic trench existed just west of the western boundary of Idaho, and the granites of the Idaho batholith were forming. Cretaceous heating and recrystallization also reset natural clocks based on radioactive minerals, giving the rocks a Cretaceous mask that hides their real age.

Most of the granitic rocks that appear on the geologic maps of the western edge of Idaho are actually diorite, a rock that looks much like granite except that it is darker and contains little or no quartz. These western Idaho diorites are generally somewhat older than the 70 or so million year old granites of central Idaho, typically closer to 90 million years old.

Think of these granites and diorites along the western edge of the Idaho batholith as the welding beads that fill the seam between the old western margin of North America and the former oceanic islands. Age dates on diorites that show little or no sign of deformation fall in the range between 90 and 100 million years. The lack of deformation is critical; it means that the seam between the western margin of the continent and the newly added oceanic islands had quit moving by about 90 million years ago, that the islands were then solidly docked onto the continent. If movement had continued after the diorites crystallized, they would now be dioritic gneisses with a streaky grain.

Granite and Pegmatite in the Palouse Range

The Palouse Range north and east of Moscow consists mostly of very pale granite that contains abundant quartz, more of the Idaho batholith. Watch for nice exposures of this granite here and there near the road, especially at Palouse Pass and just north of Viola.

The granite contains numerous quartz veins, fractures filled with milky white quartz. Most of the veins are hardly more than an inch across, but a few are much larger, and one contained enough gold to inspire the old White Cross Mine north of Moscow. The pale granite also contains numerous masses of pegmatite, a spectacular rock that consists essentially of the same minerals as ordinary granite, but differs in being composed of giant crystals anywhere from several inches to several feet across. Many pegmatites contain a wide variety of uncommon minerals, typically in pretty crystals that attract mineral collectors and rock hounds.

Pegmatites are fairly common in and around many, but by no means all, granites. Most are dikes that formed as the magma injected fractures, probably during its very last stages of crystallization. The pegmatites near Moscow are in metamorphosed sedimentary rocks

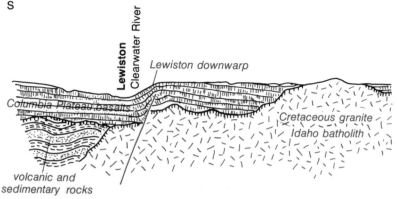

S

Lewiston
Clearwater River

Lewiston downwarp

Columbia Plateau basalts

Cretaceous granite
Idaho batholith

volcanic and
sedimentary rocks

Section just west of the line of the highway from Lewiston to the area granites and metamorphic rocks.

near the western edge of the Idaho batholith, the probable source of the magma.

The most important pegmatite mineral mined in the Palouse Range was muscovite, the white variety of mica. The giant crystals make it possible to produce mica in big sheets, instead of tiny flakes. Several mines began sporadic production of muscovite in 1888, first for isinglass windows on old fashioned ovens, later for insulating support for the wires inside vacuum tubes. Vacuum tubes are nearly as obsolete as isinglass oven windows in these days of semi-conductor devices, so the mica mines are out of business.

Some of the pegmatites contain small amounts of beryl, which typically occurs in lovely green prisms, some clear and free enough of fractures to cut as gems. A few of the beryl crystals are the blue variety called aquamarine. Rock hounds and mineral collectors have so thoroughly picked through the old mine dumps that it is now difficult to find a piece of beryl of any color.

Latah County Clay

Clay pits in the hills near Moscow began to produce high quality kaolinite for fire bricks and porcelain during the early years of the century. The white clay comes from the Latah formation, old lake beds sandwiched between flood basalt lava flows of the Columbia Plateau.

The lakes formed behind lava flows that dammed streams draining west from the northern Rocky Mountains. That happened while the region enjoyed a climate warm and humid enough to create red laterite soils and support hardwood forests similar to those that now thrive in Florida and Georgia. Kaolinite is the typical clay of laterite soils. The whiteness of the clay deposits in the Latah formation is evidence that the late Miocene lakes were acidic enough to bleach the

Moscow

Palouse River

Mission Mtn.
4324 ft.

Basement rocks

north of Potlatch. Basalt lava flows lap onto the much older

iron oxide stain from the clay, leaving kaolinite pure enough to make into porcelain.

The Columbia Plateau

Lewiston is on the Columbia Plateau, in an area where the floods of basalt backed up into the broad valley of the Clearwater River. Everywhere along the eastern margin of the Columbia Plateau, the basalt lava flows follow the valley floors east into the mountains of Idaho. Obviously, those valleys already existed when those flows erupted about 15 million years ago. We know of no clear evidence that might tell how long they had already existed before the basalt flooded them.

Just north of Lewiston, the road climbs a long hill through a tortuous series of switchbacks, a genuine adventure in winter motoring. The geologic underpinning of that hill is a sharp flexure in the basalt lava flows of the Columbia Plateau that carries flows exposed south of Lewiston to an elevation some 2000 feet higher in the hills north of town. People travelling west on Highway 12 can see wonderful exposures of the steeply tilted lava flows in the Snake River Canyon.

A spectacular display of columnar jointing in basalt near the highway about five miles south of Lewiston.

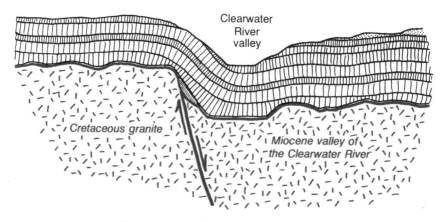

Section across the Lewiston monocline.

Such a sharp and simple fold probably means that the lava flows are draped over a fault that breaks the older rocks beneath. The Lewiston monocline is directly in line with an east-trending fault that breaks the western part of the Columbia Plateau, and probably marks its eastern continuation. The fold fades a few miles east of Lewiston, then disappears.

Palouse Hills

Where the road crosses the eastern edge of the Columbia Plateau, it passes through the billowing Palouse Hills, quite possibly the world's best wheat land. Notice that the Palouse Hills don't align into ridges, nor do the low places integrate into a drainage network. The entire vast expanse of the Palouse hills is a sea of dunes, wind blown dust. Think of the clouds of gray soil that blow up behind the machines on windy days as the wind claiming its own.

Watch in the Palouse Hills for the persistent tendency of the hills to slope gently toward the southwest, more steeply toward the northeast. Some geologists have interpreted those steep northeast slopes as the result of more rapid erosion beneath patches of snow that lingered there into the ice age summers. We agree with other geologists who interpret those steeper northeast slopes as the leading edges of dust dunes. The steepest dune slopes always face in the direction of movement.

Flood Deposits

Except for the areas near the rivers, most of Lewiston stands on the relatively flat top of a terrace that stands above the floodplain.

Wheat growing on the Palouse Hills south of Plummer.

Sediments within the upper part of that terrace were laid down from the Spokane floods, the torrents of water released again and again when the ice dam that impounded Glacial Lake Missoula at the present site of Pend Oreille Lake floated and broke. Immediately beneath the Spokane flood deposits are the ever so slightly older deposits laid down from the Bonneville flood. They are exposed in the eroded edge of the terrace, where the streets go down the slope to the level of the modern floodplain along the rivers.

When the Bonneville flood rushed down the Snake River, it backed great volumes of muddy water into the Clearwater River, reversing its direction of flow for many miles upstream. That situation must have lasted as long as thick deposits of sediment dumped from the muddy flood accumulated to record the event. It is possible to distinguish the backwater flood deposits from the ordinary deposits of the Clearwater River because they contain basalt and other rocks typical of Snake River sediments. Clearwater River sediments consist largely of pale granite brought in from the mountains to the east.

During the greatest of the Spokane floods, the water was about 600 feet deep at Lewiston, probably for two or three days each time. Those water depths backed up both the Snake and Clearwater rivers, reversing their flow far upstream from Lewiston. Each flood left its own deposit of sediment, a layer that consists of coarse gravel at the base and becomes progressively finer upward. Members of the U.S. Geological Survey have counted as many as 41 of those distinctive flood deposits in the backed up river valleys of eastern Washington and western Idaho. Evidently, Glacial Lake Missoula filled and drained at least that many times.

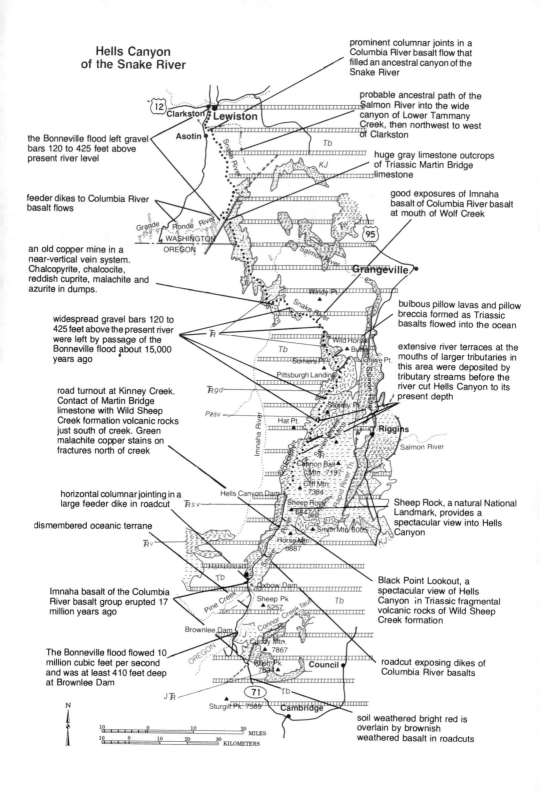

Hells Canyon
of the Snake River

prominent columnar joints in a
Columbia River basalt flow that
filled an ancestral canyon of the
Snake River

probable ancestral path of the
Salmon River into the wide
canyon of Lower Tammany
Creek, then northwest to west
of Clarkston

the Bonneville flood left gravel
bars 120 to 425 feet above
present river level

huge gray limestone outcrops
of Triassic Martin Bridge
limestone

feeder dikes to Columbia River
basalt flows

good exposures of Imnaha
basalt of Columbia River basalt
at mouth of Wolf Creek

an old copper mine in a
near-vertical vein system.
Chalcopyrite, chalcocite,
reddish cuprite, malachite and
azurite in dumps.

widespread gravel bars 120 to
425 feet above the present river
were left by passage of the
Bonneville flood about 15,000
years ago

bulbous pillow lavas and pillow
breccia formed as Triassic
basalts flowed into the ocean

extensive river terraces at the
mouths of larger tributaries in
this area were deposited by
tributary streams before the
river cut Hells Canyon to its
present depth

road turnout at Kinney Creek.
Contact of Martin Bridge
limestone with Wild Sheep
Creek formation volcanic rocks
just south of creek. Green
malachite copper stains on
fractures north of creek

horizontal columnar jointing in a
large feeder dike in roadcut

Sheep Rock, a natural National
Landmark, provides a
spectacular view into Hells
Canyon

dismembered oceanic terrane

Imnaha basalt of the Columbia
River basalt group erupted 17
million years ago

Black Point Lookout, a
spectacular view of Hells
Canyon in Triassic fragmental
volcanic rocks of Wild Sheep
Creek formation

The Bonneville flood flowed 10
million cubic feet per second
and was at least 410 feet deep
at Brownlee Dam

roadcut exposing dikes of
Columbia River basalts

soil weathered bright red is
overlain by brownish
weathered basalt in roadcuts

N

10   0   10   20 MILES
10   0   10   20   30 KILOMETERS

Hells Canyon of the Snake River, view from the rim. — Oregon State Highway Commission photo, courtesy of the Oregon Department of Geology and Mineral Industries.

HELLS CANYON

Hells Canyon, along the border between Idaho and Oregon, is the deepest canyon in the country, considerably deeper than the Grand Canyon of the Colorado River. By any standard, it is one of the scenic wonders of the continent. We find it amazing, nearly beyond imagining, that neither Idaho nor Oregon has made much effort to make it easy for people to see Hells Canyon.

Several rim view points exist in both Idaho and Oregon, but the roads that lead to them are too strenuous for most cars, and most drivers. Don't try to drive to the rim unless you are truly devoted to automotive heroics. The road from the site of the former town of Homestead, Oregon to the Hells Canyon Dam is paved and does provide a nice view of the upper part of the canyon, but not of the deepest part. Take Idaho 71 from Cambridge to Homestead. A rough but passable road leads from Whitebird, Idaho to Pittsburgh Landing, beside the river deep within the most remote part of the canyon.

Stream terraces beside the Snake River in Hells Canyon.

The only way to see the rest of Hells Canyon from the bottom up is from a raft. Several outfitters run regular raft trips through the canyon during the summer months. One of us did once float through the canyon, and can recommend it as a marvelous trip full of fun, scenery, and rocks.

A Late Miocene Spillway

Like the almost parallel Salmon River to the west, the part of the Snake River that flows nearly north along the western border of Idaho must have begun flowing across the surface of the Columbia Plateau. Basalt lava flows exposed at the rim of the canyon erupted a bit less than 15 million years ago, so the river must have eroded this deep gorge since then. We think that probably happened in two stages.

Section across Hells Canyon at Hat Point,

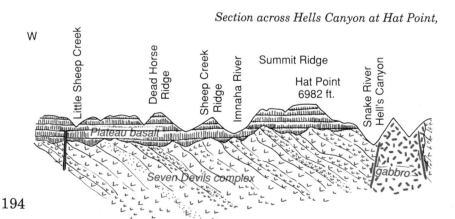

194

We suggest that the ancestors of the modern Snake and Salmon rivers began flowing in the courses that later became these canyons while the Columbia Plateau was still forming. Large lakes impounded behind basalt lava flows overflowed to establish new streams that followed the lowest available course. They began to erode valleys into the surface of the basalt plateau. That happened during the wet period that accompanied eruption of the plateau basalts. Then the climate became dry again, and those first streams may well have dried up. During the long dry period of late Miocene and Pliocene time, deep deposits of gravel accumulated in the old stream valleys. The modern Snake and Salmon rivers began to flow through those old valleys when more abundant rain began to fall about two or three million years ago. Since then, those rivers have eroded the parallel gorges that so deeply furrow this part of western Idaho.

The Canyon View

People who drive to Hells Canyon Dam and those who float the entire canyon see essentially the same assortment of geologic sights. The Blue Mountains on the Oregon side rise about 4000 to 6000 feet above the river in the deeper parts of the canyon. The Seven Devils Mountains on the Idaho side rise as much as 8000 feet above the river. Both consist of essentially similar rocks.

Most of the rocks along the canyon floor belong to the Seven Devils complex, the old islands with their fringing coral reefs that crashed into Idaho about 100 million years ago. In general, recognize the Seven Devils rocks by their layering, which is steeply tilted nearly everywhere.

Flood basalts of the Columbia Plateau line the higher canyon walls on both sides of the river, and come down to water level in several places. Look for the nearly horizontal layers of dark rock high on the canyon walls. In closer view, the dark flows distinguish themselves by their tendency to break into vertical columns.

a few miles above Pittsburgh Landing.

Seven Devils Mtns. • Martin Bridge limestone • Seven Devils complex • Riggins • Salmon River • Cougar Creek complex • Sheep Mtn. • Trick Butte 8824 ft. • granite • French Creek • gneiss • E

195

Deep inside Hells Canyon.

The last major event in forming the canyon was the Bonneville flood, which passed this way about 15,000 years ago. Although its passage did little to change the bedrock form of the canyon, the Bonneville flood did leave a few nice souvenirs.

The Old Volcanic Islands

Bedrock exposed along most of the canyon is associated in one way or another with the vagrant group of islands that docked onto the old west coast about 100 million years ago. Those rocks come in great variety.

The most abundant are the tightly folded sedimentary and volcanic rocks of the Seven Devils complex. The steeply tilted layers show that the rocks are folded. Many are considerably sheared, and metamorphosed just enough to make it difficult to know what the original rock may have been. In most cases, the originals were volcanic rocks of one kind or another, and sedimentary rocks derived from them. The Seven Devils complex is basically squashed volcanoes.

The Martin Bridge limestone makes conspicuous outcrops along the road about seven miles above Hells Canyon Dam. Watch for the thick layers of very pale gray limestone, almost white. It is full of fossils. The Martin Bridge limestone looks distinctly out of place among the other rocks of the Seven Devils complex, and it did indeed form in a completely different manner. The Martin Bridge limestone is the remains of reefs that fringed the volcanic islands.

196

The fossils in the Martin Bridge limestone very closely resemble those in European rocks of the same age, quite different from those in North American rocks. When this limestone formed, most of the Earth's inventory of continental crust was assembled into two great continents north and south of a long vanished ocean called the Tethys Sea. Europe was then at the western end of the Tethys Sea, and it is possible that these volcanic islands were somewhere near its eastern end. If so, they were then an ocean away from where you now see them.

The canyon cuts through the Cougar Creek complex of igneous and metamorphic rocks in two areas above Pittsburgh Landing. The igneous rocks, all metamorphosed, range greatly in composition from gabbros that are almost black, through dark gray diorites, to a few that are pale gray granite. The metamorphic rocks are schists and gneisses, which also range from almost black to pale gray. The metamorphism happened during Triassic time, about 230 million years ago. These rocks were then somewhere far out in the Pacific Ocean.

Columbia Plateau Basalts

Basalt flows are at river level along most of the way between Brownlee Dam and the Oxbow Dam, at the upper end of the canyon. Begin watching for them again about six miles below Pittsburgh Landing, and here and there beyond that point.

Imnaha basalt is spectacularly exposed along the river along a stretch of several miles above the Grande Ronde Canyon, watch for the imposing rows of columns standing above the river like the pipes of a giant organ. The Imnaha flows were the first series of flood basalts erupted in the Columbia Plateau. Recognize them by the glassy little crystals of green olivine that weather into specks of brownish clay, and by their tendency to weather into rounded forms.

Spheroidally weathered basalt dike, one of the feeders for the big Grande Ronde lava flows in the Columbia Plateau. This one is beside the road to Hells Canyon Dam.

197

Elsewhere, the basalt visible at close range is mostly Grande Ronde basalt, which many geologists call Yakima basalt. Those flows lack olivine, and generally make sharp columns with crisp edges. They erupted from large dikes of the Joseph dike swarm, most of which is in the area west of Hells Canyon.

Bonneville Flood

During the months while overflowing Lake Bonneville was rapidly eroding Red Rock Pass south of Pocatello, Hells Canyon ran deep with noisy torrents of muddy water. The river then carried about 1000 times as much water as it now does. The most obvious evidence of that flood's passing is in a series of giant gravel bars that rise as much as several hundred feet above the present level of the river. Watch for the streamlined hills covered with trees in the canyon floor.

As you would expect, the Bonneville flood raced through the narrower reaches of Hells Canyon and ponded in the broader reaches. Imagine the relatively quiet pools of muddy water swirling through the wide parts of the canyon, and the ferocious currents knifing viciously through the narrows.

The quiet water in the broader parts of the canyon dumped enormous amounts of sediment. After the flood passed, the river entrenched itself into those deposits, leaving remnants of them as broad terraces. Watch in the wider parts of the canyon for the flat benches along the canyon walls. The tops of those benches tell you how deep the sediment was on the canyon floor. The water was much deeper.

As elsewhere along the Snake River, the Bonneville flood left giant gravel bars in the floor of Hells Canyon. They make streamlined hills several hundred feet high, and are generally a mile or so long. Watch for them in the areas just below narrow reaches of the canyon. One of the best giant bars is beside the road to Hells Canyon Dam, where a roadside sign identifies it as the site of an early farm.

Idaho 3, Idaho 97
Lewiston — Coeur d'Alene
114 miles

The route is near the western edge of the Northern Rockies, where the flood basalts of the Columbia Plateau backed up into the old stream valleys in the western edge of the mountains. Most of the hills along and near the road contain Precambrian Belt sedimentary rocks. Other hills consist of granite and metamorphic rocks. Bedrock in low areas is generally basalt lava flows that flooded the old valleys.

Hills directly north of Deary are a small volcanic pile that erupted during Tertiary time, probably about 50 million years ago when the Challis volcanic pile was erupting.

Enigmatic Gneiss and Schist

Between Bovill and Clarkia, the road crosses a big expanse of metamorphic rocks. Watch for occasional exposures of gneiss and schist, streaky and flaky looking rocks composed of relatively large mineral grains.

Metamorphic rocks in this part of Idaho have received very little attention, so it is impossible to bring much confidence to their interpretation. We can easily imagine that they could be metamorphosed Precambrian sedimentary formations, Belt rocks that were so thoroughly baked and deformed during emplacement of the Idaho batholith that they are now gneiss and schist. It is equally easy to imagine that the metamorphic rocks between Bovill and Clarkia are Precambrian metamorphic rocks, the original basement rock of the North American continent on which the Belt formations accumulated. Gneisses formed at one time look like those formed at any other time, so their appearance is no guide to their age. Radioactive age dates might not answer the question because it is possible that heat from the nearby granite batholith could have reset the radioactive clocks in older rocks.

Outcrops of granite appear west of the road in the area a few miles north of Bovill. This is Cretaceous granite, part of the western Idaho batholith that invaded the surrounding Precambrian sedimentary formations sometime between 70 and 90 million years ago.

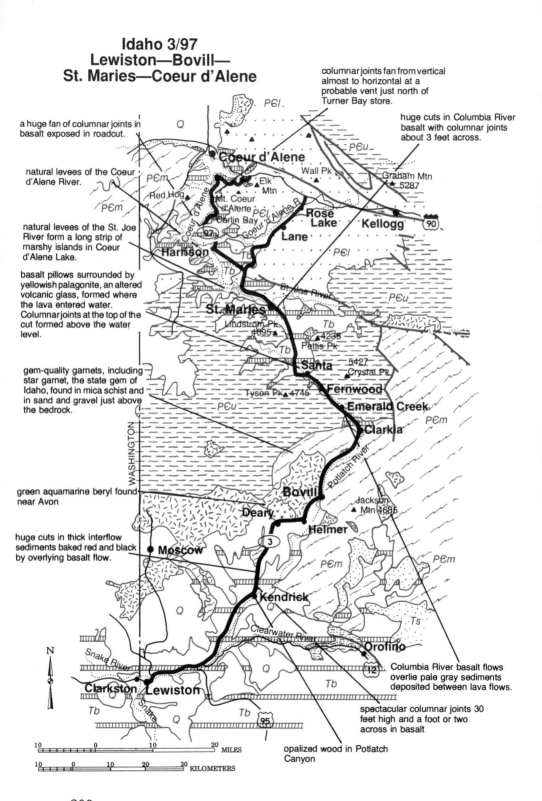

Idaho 3/97
Lewiston—Bovill—
St. Maries—Coeur d'Alene

columnar joints fan from vertical almost to horizontal at a probable vent just north of Turner Bay store.

huge cuts in Columbia River basalt with columnar joints about 3 feet across.

a huge fan of columnar joints in basalt exposed in roadcut.

natural levees of the Coeur d'Alene River.

natural levees of the St. Joe River form a long strip of marshy islands in Coeur d'Alene Lake.

basalt pillows surrounded by yellowish palagonite, an altered volcanic glass, formed where the lava entered water. Columnar joints at the top of the cut formed above the water level.

gem-quality garnets, including star garnet, the state gem of Idaho, found in mica schist and in sand and gravel just above the bedrock.

green aquamarine beryl found near Avon

huge cuts in thick interflow sediments baked red and black by overlying basalt flow.

Columbia River basalt flows overlie pale gray sediments deposited between lava flows.

spectacular columnar joints 30 feet high and a foot or two across in basalt

opalized wood in Potlatch Canyon

PЄl.

Q

PЄu

Coeur d'Alene

Wall Pk

Graham Mtn
5287

PЄm

Red Hog

Elk
Mtn

Mt. Coeur
d'Alene

Carlin Bay

PЄm

Coeur d'Alene

PЄl

Coeur d'Alene R.

Rose
Lake

Kellogg

90

Harrison

Lane

Tb

St. Joe River

PЄu

Tb

St. Maries

Lindstrom Pk
4695

Tb

4235
Pettis Pk

Tb

Santa

5427
Crystal Pk

Tyson Pk 4745

Fernwood

Emerald Creek

PЄu

PЄm

Clarkia

Potlatch River

WASHINGTON

Bovill

Jackson
Mtn 4685

Deary

Helmer

3

Moscow

PЄm

PЄm

Kendrick

Ts

Clearwater River

Orofino

N

Snake River

Q

12

Clarkston Lewiston

Tb

Q

Tb

Snake

95

Tb

Tb

| 10 | 0 | 10 | 20 | MILES |

| 10 | 0 | 10 | 20 | 30 | KILOMETERS |

200

Belt Formations

Except for several fairly small areas of Columbia Plateau basalt, all the rocks between Clarkia and Coeur d'Alene are Precambrian sedimentary formations, Belt rocks. Those south of St. Maries are mostly the younger Belt formations that tend to contain colorful red and green mudstones and sandstones in various pastel colors. Most of those rocks were laid down in shallow water or on land.

Belt rocks north of St. Maries are mostly the much older Prichard and Ravalli formations that appear to have accumulated in fairly deep water. They consist largely of somber mudstones and muddy sandstones in various shades of dark gray. Many acquire a rust stain as the iron pyrite in them begins to weather, a warm wash of brown over the basic dark gray of the fresh rock. Especially good exposures of those rocks exist along the road near the north end of Coeur d'Alene Lake, around Beauty Bay.

Plateau Basalt

Watch for long palisades of basalt columns here and there along the east side of Coeur d'Alene Lake — between the south end of the lake and the area around St. Maries, at Clarkia, and along the Potlatch River from a few miles north of Kendrick south.

An especially splendid road cut about three miles north of Santa exposes the record of a basalt flow that poured into a lake. It contains a very nice mess of dark basalt pillows scattered through a matrix of yellowish brown palagonite, the clay that forms as molten basalt reacts with water. A regular row of basalt columns at the top of the exposure identifies the part of the lava flow that stood above the water.

Between Lewiston and Kendrick, the road follows the Clearwater and Potlatch rivers, travelling on basalt lava flows that flooded into the valley about 15 million years ago, part of the Columbia Plateau. The road continues on basalt lava flows most of the way between Kendrick and Deary, but a cover of younger sediments makes them less visible near Deary. The long stretch of road between Santa and Coeur d'Alene is also on basalt, in this area lava flows that flooded into the valley of the St. Joe River.

Throughout the western part of central Idaho, the big stream valleys tend to appear on the geologic map as narrow avenues of basalt winding through hills eroded in the much older rocks. That pattern is clear evidence that those valleys already existed in essentially their present form when the lava flooded them some 15 million or so years ago. The stream valleys were in the same places then as now, and the same hills that stood above the red hot floods of lava still rise above the basalt flows in the stream valleys. Fifteen million years of erosion have neither cleaned the basalt out of the valleys nor levelled the hills.

Palouse Hills

The silt dunes of the Palouse Hills cover most of the basalt, making a lovely rolling landscape covered with prosperous farms. But this landscape is as peculiar as it is picturesque. In fact, it is bizarre. Look closely; the Palouse Hills are not a normal erosional landscape. Notice that the hollows in the Palouse country generally contain no streams and do not connect to make continuous valleys. Neither do the hills connect to make long ridges. The landscape consists of isolated humps and hollows, more or less randomly arranged. The Palouse Hills are simply a sea of wind dunes, a relic of a time when the climate was extremely dry. Their age is actually unknown, but we think the Palouse Hills probably formed before the ice ages.

Notice that the steepest slopes in the Palouse country tend to be on the sides of the hills that face northeast. Dunes invariably form their steepest slopes on their downwind sides, in the direction of their movement. Evidently, strong southwest winds drove the Palouse dunes northeast. The migrating dunes marched across an older erosional landscape carved into the basalt flows, filling the valleys and burying most of the hills. Widely scattered exposures of basalt in the Palouse Hills are probably the tops of old hills that the wind blown silt did not quite bury. They tend to support groves of trees and, in many cases, road metal pits.

Farmers who till the Palouse Hills tend not to worry much about erosion because the silt deposits are so deep. That is probably a mistake. Windblown silt is extremely vulnerable to wind erosion — what the wind gave it can also take away. Enormous clouds of dust often rise from working farm equipment, and commonly drift downwind as far as western Montana. Closely-spaced rills and gullies on many bare slopes show that the soil is also vulnerable to rainsplash and surface runoff erosion. Some conservationists maintain that the region's loss of soil to erosion considerably exceeds its wheat production, measured in tons.

The Natural Levees of the St. Joe River

Chatcolet and Benewah lakes are actually two shallow bays at the south end of Coeur d'Alene Lake. Both are rapidly silting up, and they are terribly weedy because sewage pollution provides them with an excess of fertilizer nutrients. The natural levees of the St. Joe River, which separate the two bays, are the big geologic attraction.

The natural levees of the St. Joe River in the south end of Coeur d'Alene Lake.

Many streams build natural levees, simply deposits of flood sediment that gradually raise the banks a bit above the level of the adjacent floodplain. Most natural levees are such low and subtle lifts in the landscape that you rarely notice them. You need a surveying level to detect them.

The natural levees of the St. Joe River continue from the floodplain, where no one notices them, out into the south end of Coeur d'Alene Lake, where they become obvious as a pair of low embankments that carry the river far out into the lake. They probably formed in the normal inconspicuous way on the floodplain, then came to everyone's attention after a dam raised the lake level several feet, just enough to inundate the low parts of the old floodplain while leaving the slightly higher crests of the natural levees dry.

Garnets: Raw Material for Sandpaper and Gems

Although Emerald Creek is certifiably free of emeralds, it does contain large placer deposits of garnets that weathered out of metamorphic rocks in the border zone of the Idaho batholith. Mines near Fernwood have produced garnets for years, thousands of tons of them for abrasives, handfuls for gems.

Garnet is a common mineral in many kinds of metamorphic rocks, as well as in a few igneous rocks. Most garnets are red, quite a few are bright red. Given half a chance, garnets crystallize into beautifully geometric forms. The mineral is easy to like.

Garnet is considerably denser than most common minerals, so it tends to concentrate in placers, along with such other heavy minerals as magnetite, zircon, and gold. Look with a magnifying glass at almost any streak of black sand in a central Idaho stream bed, and you will probably see sparkles of red garnet.

Garnet makes a good abrasive because it is hard and crushes into angular pieces with razor edges that finish wood faster and smoother than most other abrasives. Garnet is also much used in sand blasting because the grains are heavier than those of most minerals, so they hit harder.

Pretty as they are, most garnets are too fractured to cut and polish, too full of impurities or too intensely colored to make attractive gem stones. Those rare few transparent grains that are free of fractures and considerably paler than most garnets make beautiful deep red gem stones. Several of the placers produce such garnets, including the beautiful star garnets sold in gem shops all over Idaho.

Star garnets contain mineral impurities that grow as fibers finer than hairs along certain preferred directions within the crystal. Internal reflections from those geometrically arranged fibers make the ghostly star within the stone. Those stones are normally cut as cabochons to better display the star.

Most outcrops within ten miles of Boise expose black basalt lava flows on the surface of the Snake River Plain. They erupted as Basin and Range faults began to break this region. Outcrops farther north and east expose the much older and much paler rocks of the northern Rocky Mountains. Bedrock exposed along the road between Arrow Rock Reservoir and Stanley is mostly granite cut in many places by andesite and basalt dikes, which formed as magma injected fractures. Watch for road cuts in sparkling white or pale gray granite in which the dikes appear as nearly vertical dark stripes.

Granite and Granite

Virtually all the granite beside the road belongs to the southern part of the Idaho batholith, the Atlanta batholith. It formed during late Cretaceous time, between 70 and 90 million years ago, in round numbers. The granite is pale gray on fresh exposures, and in some places very slightly streaky. Weathered surfaces become pinkish brown, an iron oxide stain that develops as the black minerals weather. Even that color becomes difficult to see in places where lichens cover weathered outcrops with a richly variegated carpet of black, green, and yellow.

The much younger Sawtooth batholith, about 50 million years old, forms a large part of the Sawtooth Range. It reaches the road in one small area about 20 miles northeast of Lowman. Granite of the Sawtooth batholith differs distinctively from that of the Idaho batholith, most obviously in being pink instead of gray in fresh exposures. It also contains distinctly larger crystals than the older granite, and is everywhere massive, with no sign of streakiness. In all those respects the granite of the Sawtooth batholith resembles the other 50 million year old granites. The granites of central Idaho do indeed betray their age by their appearance.

Swarms of Dikes

The geologic map shows swarms of dikes in the area north of the road; dense patterns of lines that only hint at the extravagant reality. In fact, thousands of dikes cut that area. Most trend generally northeast, as does the swarm, which continues right through central Idaho and southwestern Montana — generally from Boise to Butte.

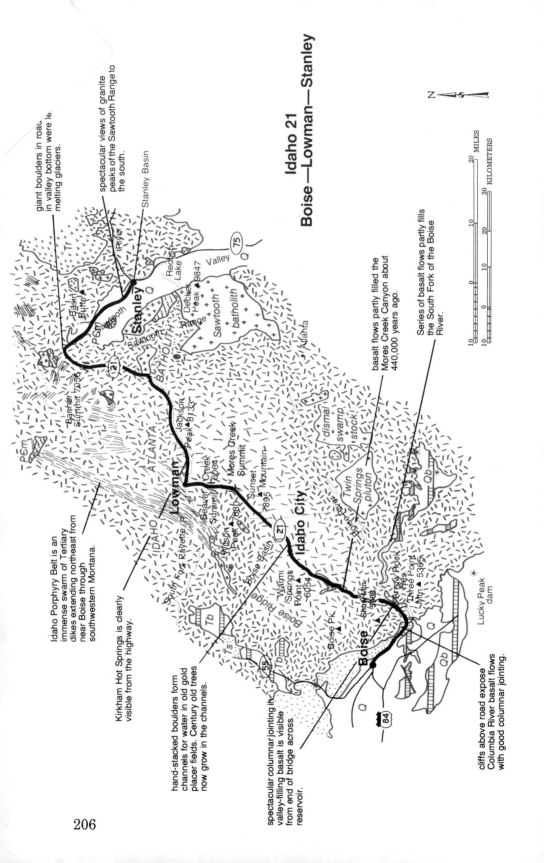

Idaho 21
Boise—Lowman—Stanley

N

MILES
KILOMETERS

giant boulders in road
in valley bottom were le
melting glaciers.

spectacular views of granite
peaks of the Sawtooth Range to
the south.

Stanley Basin

Redfish
Lake

Valley   75

Basin
Butte

Stanley

Sawtooth
Summit 7056

Sawtooth
Range

Decker
Peak 9847

Sawtooth
+ batholith

Atlanta

SAWTOOTH
BATHOLITH

Jackson
Peak 8733

Mores Creek
Summit

Sunset
Mountain
7895

ATLANTA

Lowman

IDAHO

South Fork Payette R

Beaver
Creek
Summit 7064

Wilson
Peak 880

Boise Basin

Idaho City

21

dismal
swamp

Twin
Springs
stock

Springs
pluton

Boise River

basalt flows partly filled the
Mores Creek Canyon about
440,000 years ago.

Series of basalt flows partly fills
the South Fork of the Boise
River.

Idaho Porphyry Belt is an
immense swarm of Tertiary
dikes extending northeast from
near Boise through
southwestern Montana.

Kirkham Hot Springs is clearly
visible from the highway.

hand-stacked boulders form
channels for water in old gold
placer fields. Century old trees
now grow in the channels.

spectacular columnar jointing in
valley-filling basalt is visible
from end of bridge across
reservoir.

Tb

Ts

Tb

55

Boise Ridge

Warm
Springs
Point
6054

Boise Pk

Boise

Shafer Mtn
5968

Arrow Rock
Res.

Three Point
Mtn 5365

Lucky Peak
dam

cliffs above road expose
Columbia River basalt flows
with good columnar jointing.

Q

Qb

Q

Qb

84

PЄm

Tr

Tr

PЄm

206

Plenty of good age dates leave no doubt that the dike swarm is about 50 million years old, the same age as the Challis volcanic rocks and many granite intrusions in central Idaho. Furthermore, the compositions of the dikes, the Challis volcanic rocks, and the granite intrusions all fall within the same fairly narrow range. Such similar rocks formed in the same area at the same time must be related. We think the magma that became the dikes, the granite intrusions, and the Challis volcanic pile formed through partial melting of the Idaho batholith, which was then only about 20 million years old, and still hot inside.

Gold in the Boise Basin

Prospectors found gold in the creeks of the Boise Basin area around Idaho City in 1862. That discovery promptly launched a mining spree that was long past its peak by 1869, but limped on until about 1952.

As usual, the first and easiest pickings in the Boise Basin went to the miners hand working the gravels in the stream beds; they had the least need for large investment of capital and labor before they could get into production. In all likelihood, those were the miners that made the greatest return on their investment.

After those early independents had done what they could, companies of miners dug long ditches to bring water to work the higher gravels hydraulically. That is a drastic process that involves washing the slope down with jets of water shot under high pressure through nozzles called giants, which are so big that you can easily mistake one for an old cannon. The hydraulic workings left deep scars, now largely overgrown with trees, on long stretches of hillslope. They also left gravel tailings on the valley floor.

These neatly stacked cobbles lining an old ditch near Idaho City mark old diggings that probably date from the earliest workings, perhaps about 1863.

207

Finally, beginning in 1898, big dredges that cost millions of dollars methodically turned the floodplains upside down as they extracted almost everything that remained. The dredges also reworked most of the tailings left from the earlier operations.

Long windrows of gravel, mostly dredge tailings, in the floodplains tell the story. Watch for several miles of placer mine spoils along the road and all across the valley floor, especially near Idaho City. Those unsightly heaps of gravel will probably continue to tell the story for many thousands of years before natural processes can begin to restore the floodplains to anything resembling their natural condition.

The first discoveries of lode deposits, gold bearing veins in the bedrock, came sometime during 1864. That stage of prospecting rarely takes long. Bedrock mining went through the usual succession of stages: early use of primitive arrastras that used mule power to grind the ore; later use of expensive stamp mills that depended on water or steam power; finally, installation of cyanide mills that extract gold chemically.

Widely accepted estimates place total production from the Boise Basin placers at 2,300,000 ounces. The hardrock mines produced much less, about 425,000 ounces, most of it from the area around Quartzburg. Those amounts make the Boise Basin the most productive gold mining district in Idaho, beyond question.

The Atlanta Mining District

The thoroughly derelict town of Atlanta, about 40 miles east of the highway, was for a few hectic years the center of one of the most important mining districts in Idaho. The glory years were as quick in passing as they were slow in coming.

Prospectors found gold near Atlanta in 1863, placer deposits in the stream gravels and a truly fabulous quartz vein in the bedrock. As it happened, the placer deposits were lean and the bedrock ore contained a larger proportion of silver than the early milling methods could handle. One new method after another failed to solve the problems. Furthermore, most of the ore body proved to be a large mass of low grade rock that would require mining on a much larger scale than the operators of the last century could manage — they specialized in quick profits from high grade ore.

Finally, in 1932, a new mill using a novel process that recovered both silver and gold made Atlanta the biggest gold producer in the state until 1936. Another mine produced more gold and some antimony between 1947 and 1953. Most estimates have Atlanta producing a total of some 16 million dollars worth of gold and silver, the last of it in 1953.

Black Sand Placers

Bear Valley Creek, across the divide north of Lowman, was the site of large dredging operations to mine the black sand that gold miners generally consider a nuisance. Black sands are placer deposits. They consist of an assortment of minerals that tend to concentrate together because they are denser than most minerals. Black sands typically contain such minerals as magnetite, garnet, zircon, and gold, but vary greatly from one stream to another, depending upon the kinds of rocks exposed in the watershed.

In addition to the usual content of garnet and magnetite, the black sands of Bear Valley Creek contain large proportions of some rather unusual minerals. Those include monazite, a rare earth mineral; the uranium mineral euxenite; columbite, which contains niobium and columbium; and zircon, the only source of zirconium. That assemblage of heavy minerals was mined mainly as a source of columbium, niobium, zirconium, and tantalum, exotic elements with many exotic space age uses.

Basalt

Between Idaho City and the Lucky Peak Dam, the highway passes numerous exposures of basalt, well above the road near Idaho City, at or below road level closer to the dam. Watch for the black rocks with their neat rows of nifty vertical columns. Evidently a lava flow once covered the floor of this valley, then the river cut through it into the pale granite beneath. It seems reasonable to suggest that the flow erupted a few million years ago, as the eastern edge of the Basin and Range passed this way.

The palisade of vertical columns is a basalt flow that rests on creek gravel near Idaho City. —U.S.Geological Survey photo by C.P. Ross

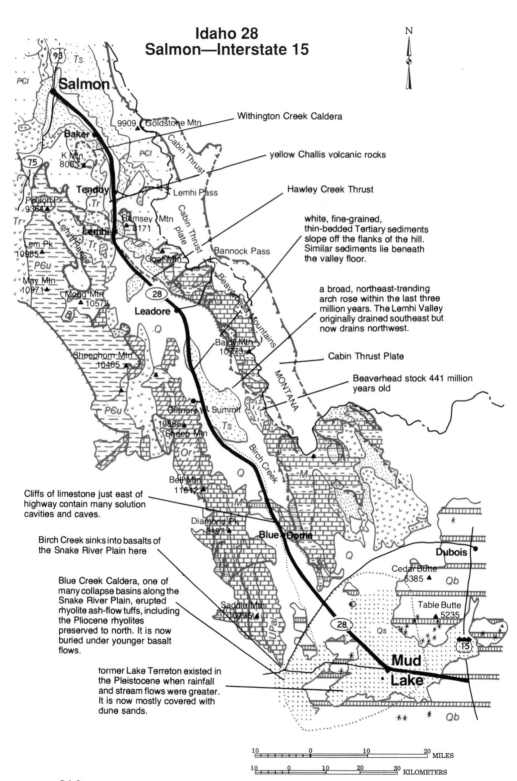

Idaho 28
Salmon—Interstate 15

N

Withington Creek Caldera

yellow Challis volcanic rocks

Hawley Creek Thrust

white, fine-grained, thin-bedded Tertiary sediments slope off the flanks of the hill. Similar sediments lie beneath the valley floor.

a broad, northeast-trending arch rose within the last three million years. The Lemhi Valley originally drained southeast but now drains northwest.

Cabin Thrust Plate

Beaverhead stock 441 million years old

Cliffs of limestone just east of highway contain many solution cavities and caves.

Birch Creek sinks into basalts of the Snake River Plain here

Blue Creek Caldera, one of many collapse basins along the Snake River Plain, erupted rhyolite ash-flow tuffs, including the Pliocene rhyolites preserved to north. It is now buried under younger basalt flows.

former Lake Terreton existed in the Pleistocene when rainfall and stream flows were greater. It is now mostly covered with dune sands.

Salmon

Baker

K Mtn 8063

Tendoy

Poison Pk 9354

Lemhi

Lem Pk 10985

May Mtn 10971

Mogg Mtn 10573

Leadore

Sheephorn Mtn 10465

Gilmore Summit 10885
Sheep Mtn

Bell Mtn 11612

Diamond Pk 9197

Saddle Mtn 10795

Goldstone Mtn 9909

Lemhi Pass

Ramsey Mtn 9171

Goat Mtn

Bannock Pass

Baldy Mtn 10773

Blue Dome

Cabin Thrust

Cabin Thrust Plate

Lemhi Range

Beaverhead Mountains

MONTANA

Birch Creek

Mud Lake

Dubois

Cedar Butte 5385

Table Butte 5235

```
10        0        10        20 MILES
10    0    10    20    30 KILOMETERS
```

210

Basalt flow east of Highway 28 at Lonepine.

Idaho 28:
Salmon — Interstate 15
133 miles

Highway 28 follows the semi-arid Lemhi Valley, a broad corridor of low country that separates the Lemhi Range on the southwest from the Beaverhead Mountains to the northeast. The Lemhi River flows north through the northern part of the basin to join the Salmon River; Birch Creek drains its southern part into a sink in the absorbent lavas of the Snake River Plain. Despite the two drainages, it is all one structural basin.

Except for a basalt lava flow a few miles north of Blue Dome, all of the rocks exposed along the road and in the valley floor are basin-fill sediments and stream deposits. The older bedrock is all in the mountains, well away from the road.

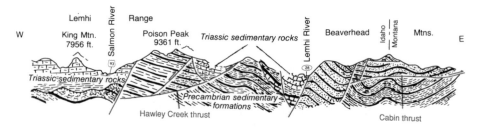

Section across the Lemhi and Beaverhead Mountains just north of Lemhi. The steeply dipping Basin and Range faults cut across the much older thrust faults.

The Cabin and Hawley Creek Thrust Plates

The Lemhi and Beaverhead mountains contain a rich assortment of folded and faulted Precambrian, Paleozoic, and Mesozoic sedimentary rocks, all deformed while the northern Rocky Mountains rose during late Cretaceous time. Large slices of rock, thrust plates, moved east as they uncovered the granite of the southern part of the Idaho batholith, then came to rest in this part of Idaho. Those movements left the thrust slices stacked on each other to form part of the overthrust belt.

Between Lemhi and Blue Dome, the road is on the Hawley Creek thrust plate, the slab that moved east on the Hawley Creek thrust fault. That fault comes to the surface in the Beaverhead Mountains, where early Paleozoic rocks more than 500 million years old lie on late Paleozoic rocks about 300 million years old. The road crosses that Hawley Creek fault about two miles south of Lemhi, where it moved Precambrian sedimentary formations a billion or more years old onto lower Paleozoic rocks.

The road between Salmon and Lemhi is on the Cabin thrust plate, a slab that moved east onto the Hawley Creek thrust plate, and stacked on top of it. The outcrop of the Cabin thrust fault lies along the east side of the Beaverhead Mountains, in Montana.

Three Fault Blocks

The Lemhi Valley is a block of the northern Rocky Mountain overthrust belt that dropped along Basin and Range faults during the last several million years, while the mountain blocks on either side rose. Those movements continue to raise the mountains and drop the valleys. You see those faults in the remarkably straight mountain fronts on either side of the Lemhi Valley.

The Lemhi Valley accumulated deep deposits of desert valley-fill sediments during Pliocene time, when the climate was so dry that no streams drained the region. As Birch Creek and the Lemhi River began to flow on those deposits when the ice ages began, they formed the low drainage divide that now gives the opposite ends of the basin their different names. The modern streams have now carved their erosional valleys deeply into those deposits, leaving remnants of the original basin-fill surface on the high benches that line the valley walls.

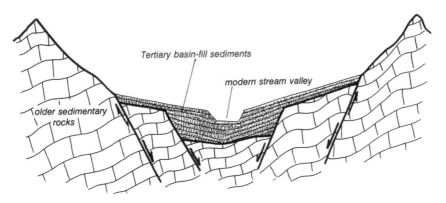

A schematic cross-section showing how the modern streams cut their valleys into the basin-fill sediments.

The Cretaceous Beaverhead gravel that forms large patches of the Lemhi and Beaverhead ranges is as curious as any rock formation we know. It was deposited while crustal movements shaped the Rocky Mountains, then folded while it was still loose gravel. In many places, the stresses of deformation jammed pebbles into each other, mashing and shearing them as though they were a stepped-on basket of hard boiled eggs. The tortured pebbles have since welded themselves together to make some of the most strangely distorted rocks imaginable, marvelous paper weights.

Challis volcanic rocks covered large areas of the older rocks in the Lemhi Range and lesser areas of the Beaverhead Mountains during Eocene time, about 50 million years ago. They consist mostly of pale rhyolite ash. You can recognize from a distance the parts of the ranges still covered by those volcanic rocks by watching for areas where closely spaced small streams cut the mountains into an intricately dissected landscape. The older rocks tend to erode into broader, more rounded hills.

Pebbles of the Beaverhead Conglomerate weather out of a hillside.
—U.S.Geological Survey photo by C.P. Ross

Younger volcanic rocks, rhyolite ash and basalt lava flows, erupted within the last few million years. Some, such as the basalt flow near Blue Dome are clearly associated with Basin and Range faulting. The younger rhyolite is ash that blew in from the Snake River Plain. The southern ends of the Lemhi and Beaverhead ranges vanish into the Blue Creek caldera, one of the big rhyolite volcanoes that built the Snake River Plain.

Blue Creek Caldera

The southern end of the route crosses the Blue Creek caldera, an earlier version of the modern Yellowstone volcano. It erupted enormous volumes of rhyolite ash that boiled out of the volcano in clouds of red hot steam, then poured still half-molten across the surrounding countryside to fuse into more or less solid rhyolite. After the last rhyolite cataclysm, volcanic eruptions associated with Basin and Range faulting spread a veneer of basalt lava flows across the ruins of the old volcano. That is the bedrock you now see on most of the surface. In fact, the younger volcanic rocks so thoroughly fill the old caldera crater that its outlines are only dimly visible on the geologic map, not at all on the ground.

Thorium and Rare Earth Elements at Lemhi Pass

Veins that cut Precambrian sedimentary rocks at Lemhi Pass on the Montana line east of Tendoy contain large amounts of thorium and rare earth minerals, as well as small concentrations of gold, lead, zinc, and copper. According to some estimates, which are probably conservative, the reserves of thorium at Lemhi Pass are the largest known to exist in the United States.

Thorium is an interesting radioactive element with almost no industrial use. If a significant market for thorium should ever develop, the deposits at Lemhi Pass might someday support mines, probably open pits. Rare earth elements do have their uses, so it is possible that mines may someday produce them at Lemhi Pass, even if no demand for thorium ever materializes.

Glaciation

The higher parts of the Lemhi and Beaverhead ranges caught enough snow out of ice age storm clouds to support large glaciers. Watch for the ragged high peaks that drop steeply off into deeply gouged valleys, infallibly the signs of glacial sculpture.

The earlier and greater of the two major glaciations that affected Idaho left moraines near the mouths of the larger valleys, showing that glacial ice almost reached the floor of the Lemhi Valley. The much smaller glaciers of the most recent ice age left their moraines farther up the valleys.

Lake Terreton

Mud Lake is the shrivelled remnant of a much larger body of water, Lake Terreton, which existed during the ice ages when the climate was considerably wetter than what we know today. Lake Terreton flooded a large part of the low center of the old Blue Creek caldera. Basalt shield volcanoes form the rim of low hills that confine the basin.

As Lake Terreton shrank in the drier post-glacial climate, it became a desert playa that flooded temporarily after the occasional heavy rain, then dried up in the sunny days that followed. A large area of mud deposits laid down during those fillings records the story. Playa lake sediments commonly include sand dunes such as those north of the road west of Mud Lake.

steep cliffs of Columbia River basalt in Ponderosa State Park overlook Payette Lake

North Loon Mtn. 8280 ▲

"block stream," a large slide of huge boulders of granite just east of Summit Lake on Lick Creek Road

Slick Rock, a shear glaciated cliff above the west side of Lick Creek Road

Payette Lake

New Meadows

McCall

55

OREGON

Snake River

Tb

Donnelly

Q

Council

Council Mtn. 8126 ▲

Columbia River basalt flows have been tilted about 30 degrees westward

95

Eagle Nest 7646 ▲

Idaho Atlanta Batholith

Indian Mtn. 7263 ▲

Lookout Peak 7813 ▲

Cascade

moraines here were deposited by glaciers that flowed down the valley of Payette Lake

Cascade Res.

East Mtn. 7752 ▲

Kg

Tb

Tripod Pk. 8086 ▲

Rattlesnake Point 7250 ▲

West Mountains

Smiths Ferry

Q

North Fork Payette River

Middle Fk. Payette River

Banks

Ts

455 ▲

Timber Butte

Hawley Mtn. 7501 ▲

Gardena

Tb

Boise Basin

Payette River

Q

Horseshoe Bend

Idaho Atlanta Batholith

Emmett

Ts

Kg

Imnaha basalt of the Columbia River basalt group erupted about 17 million years ago

Shafer Butte 7582 ▲

highway cut through Tertiary sediments continually slides and slumps

Boise Peak ▲

21

beige Miocene sands eroded from the Idaho batholith

Q

Boise

84

Boise River

Nampa

Idaho 55
Boise—New Meadows

Qb

10   0   10   20   30   KILOMETERS

N

216

Idaho 55:
Boise — New Meadows
120 miles

Boise is at the northern edge of the Snake River Plain, about where it meets the Columbia Plateau. New Meadows is near the eastern edge of the Columbia Plateau. Exposures of basalt lava flows appear near both ends of the route, although younger sediments make them hard to see.

Between McCall and Horseshoe Bend, the road follows the lovely Payette River through the late Cretaceous granites of the western part of the Atlanta batholith, all between 70 and 90 million years old. Watch for the picturesque bouldery outcrops. Between Horseshoe Bend and the main part of the Snake River Plain a few miles north of Boise, the road crosses basalt that flooded an older valley eroded in the granite.

Forested mountains that appear here and there on the skyline ten or more miles west of the highway are eroded into the Seven Devils complex. Those exotic rocks originally formed in oceanic islands somewhere far out in the Pacific, then joined the old western margin of North America about 100 million years ago. The experience of driving past former South Pacific islands now grounded in Idaho and covered with pine trees may not quite match that of visiting their modern equivalents, but this road doesn't go to Tahiti.

Section across the line of the road between McCall and Cascade showing the Atlanta batholith invading metamorphic rocks. Plateau basalts flood Long Valley.

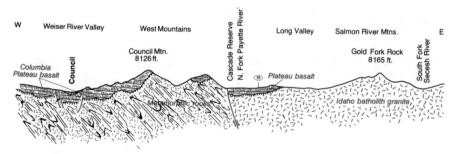

217

The Atlanta Batholith

Although the Bitterroot and Atlanta portions of the Idaho batholith underlie most of central Idaho, the region is so short of roads that you rarely see much of either. The long drive between Cascade and Horseshoe Bend provides one of the best views from a paved road of the Atlanta batholith. It is an awesome expanse of granite.

Like granites nearly everywhere, this one tends to weather into highly characteristic rounded boulders that make the rock easy to recognize at a glance. In this region, granites generally acquire a rusty brown stain of iron oxide as they weather, and a crust of black lichens. Between them, the iron oxide and the lichens give the boulders a dark brown color with just a hint of pink. So it is a surprise to hammer off a chip and find that the sparkling fresh rock is pale gray, almost white. Look at the pale cobbles and boulders in the bed of the Payette River.

A close look at a fresh chip shows that the most abundant mineral in the granite is feldspar, most of it milky white. Any that looks even slightly pink is the potassium feldspar orthoclase. Feldspar grains tend to have flat surfaces, and many show blocky outlines. Quartz forms rounded grains that generally look gray because you see through the glassy mineral into the shadowed place it fills. Black flakes are biotite mica. This granite is nearly massive; it shows almost no hint of the streaky grain that develops if the mineral grains line up parallel to each other, as they do in gneiss.

The boulders that litter so much of the ground surface form in two stages: First, the rock weathers along the widely spaced fractures to form soil that encloses unweathered rock in the cores of the spaces between fractures. The core stones become rounded because weathering preferentially attacks sharp edges and corners. Next, the soil erodes, leaving the unweathered and rounded core stones on the surface. That probably happens when a long interval of dry climate reduces the plant cover enough to expose the soil to rainsplash erosion. An intense forest fire might have the same effect. So, the slopes littered with rounded boulders have simply lost most of their original upholstery of soil. The boulders were rounded where they are, without moving.

Columbia Plateau Basalt

Long stretches of the road pass one somber exposure after another of dark brown and black basalt, part of the Columbia Plateau. It is always easy to recognize a basalt lava flow in a road cut or cliff exposure. Watch for the stockades of vertical black columns with their

A typical palisade of basalt columns.

flat sides and crisply defined edges. Nothing else looks quite like them. It is generally easy to interpret the columns.

Simply remember that the columns form as the lava flow crystallizes, invariably in a direction at right angles to the surface of the lava flow. In the simplest case, the flow pours across fairly level ground and the columns form vertically, at right angles to both the ground surface and the top of the flow. If, as often happens, the flow fills a stream valley, the columns will develop in a fan pattern, each meeting the old valley wall at something close to a right angle. Watch for those wild fans; they are fairly common along many Idaho roads.

We tend to assume that each lava flow has one row of columns, and that we can count flows by simply ticking off the number of successive palisades of columns, each forming its own cliff exposure. That simple approach seems to work well enough if the lava flows are thin, but thick flows, those more than 100 or so feet from bottom to top, tend to fracture into two sets of regular columns at the top and base of the flow. The irregularly fractured zone between the regular collonades is called a hackly entablature. Early geologists in the Columbia Plateau consistently attempted to count lava flows by ticking off the number of ledges that exposed rows of columns, and almost as consistently counted twice as many flows as actually exist.

Bold columns of basalt. Most have five or six sides.

You may read somewhere that basalt columns have six sides arranged in the outlines of more or less regular hexagons. Not so. In fact, basalt columns may have anything from four to seven sides, with five being much the commonest number. Pause for a few minutes to count column sides in some nice road cut.

Yellowpine Mines

The side road east from McCall eventually winds its way to the old mining districts around Yellowpine and Stibnite. Mines in that area produced enormous amounts of tungsten, antimony, gold, and mercury, along with lesser amounts of other metals. The ore is in veins that fill generally north trending fractures in the granite of the Idaho batholith.

Antimony is a metal of many uses in making alloys, pigments, batteries, and quite a variety of other products. In the Yellowpine Mine, it occurs in a spectacular mineral called stibnite that crystallizes into marvelously shiny black needles. Production there began in

1932, and continues off and on as the price of antimony fluctuates and exploration projects reveal new veins.

Tungsten has a variety of uses, most familiarly as the metal that makes the filaments in light bulbs, equally importantly as an essential ingredient in high speed tool steels, among other useful products that an industrial economy must have. Large reserves of tungsten were discovered at Yellowpine in 1941, just in time for the Second World War. They were gone by the end of 1944, after having provided about 40 percent of the national supply during those difficult years.

Most of the tungsten in the Yellowpine Mine was in a nondescript white or slightly pink mineral called scheelite. It would be one of the easiest of all ore minerals to overlook if it did not fluoresce so strongly under ultraviolet light. Prospectors searching for scheelite sally forth in the night armed with a black light.

The same veins that provided all the antimony and tungsten also contained enough gold to make the Yellowpine Mine the biggest producer in the state between 1946 and 1951. The gravels in streams draining the area contain no placer deposits worth mentioning because the gold does not occur as the native metal. Instead, it is an impurity in pyrite and in a closely related mineral called arsenopyrite, which also contains arsenic. Pyrite and arsenopyrite both crystallize into shiny metallic cubes, but you can distinguish them by their colors. Pyrite is pale yellow; arsenopyrite is silvery white.

The Hermes Mine near Yellowpine produced a large part of all the mercury that ever came out of Idaho, second only to the amount that came from the Idaho-Almaden Mine near Weiser. The deposit was discovered in 1902 in a small area of metamorphic rocks within the Idaho batholith. Like most mercury deposits, these probably precipitated from hot water at very shallow depth beneath a hot spring.

The Hermes Mine went into full operation in 1942, and by 1948 had produced 10,700 flasks of mercury, 76 pounds to a flask. The mine fed two rotary furnaces, each capable of roasting a daily ration of 75 tons of ore. After 1948, the property went into rapid decline, but continued in sporadic production until about 1960.

An Oversized Valley

Between Cascade and McCall, the road follows a valley much too large to have been eroded by the small streams that now wander here and there, almost lost, on its broad floor. It is probably a block of the earth's crust that is dropping along faults, a northern outpost of the Basin and Range deformation so prominent farther south. The valley has the right northerly trend and its rather straight sides look like they might be fault scarps.

The floor of the valley contains a deep fill of glacial debris, outwash swept into it by meltwater coming in from the mountains to the east. Payette Lake north of McCall fills a natural basin impounded by the glacial outwash. A dam impounds Cascade Reservoir in the southern end of the valley.

A series of north-south ridges that the highway crosses between New Meadows and McCall displays even more clearly a series of Basin and Range fault blocks. Each block consists of Triassic-Jurassic igneous rocks capped by Columbia Plateau basalt flows tilted about 20 degrees to the west.

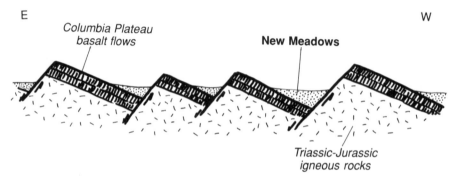

Tilted Basin and Range fault blocks south of the highway between New Meadows and McCall.

Ice age glaciers carved the granite of the Sawtooth Range into this sea of jagged peaks that rise above glacially gouged valleys.
— U.S. Geological Survey photo by E.Day

Idaho 75:
Shoshone — Challis
171 miles

This beautiful route crosses such a fascinating variety of rocks that it is impossible to summarize them in a few words. We will begin with a brief tour of the geologic landscape to introduce the rocks along various stretches of the road, then go on to the details.

Between Shoshone and the area near Bellevue, the highway crosses the northern part of the Snake River Plain. Rocks near that part of the road are basalt lava flows, part of the crust that covers the much greater volume of rhyolite beneath. Rhyolite does surface in some of the hills near the road.

Between Bellevue and Galena Summit, the highway follows the Wood River through the geologically complex mountains of central Idaho. The older rocks are Paleozoic sedimentary formations that were tightly folded, much broken along faults, and extensively intruded by granite of the Idaho batholith during late Cretaceous time,

223

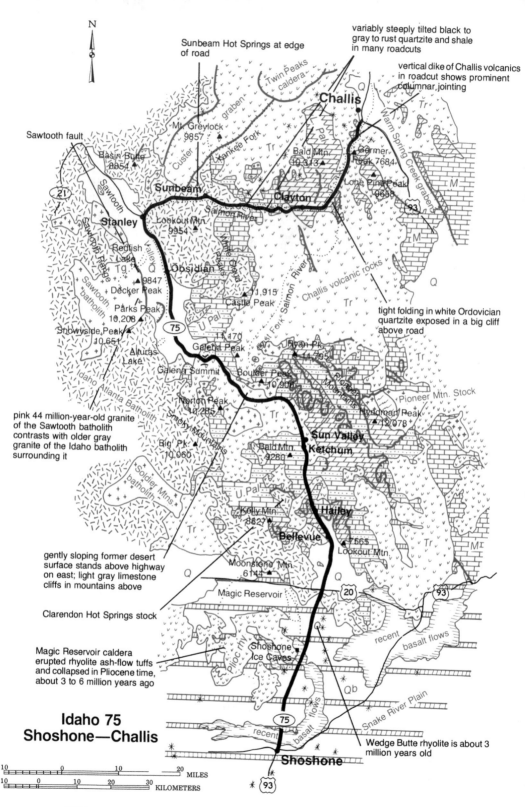

N

Sunbeam Hot Springs at edge of road

variably steeply tilted black to gray to rust quartzite and shale in many roadcuts

vertical dike of Challis volcanics in roadcut shows prominent columnar jointing

Twin Peaks caldera

Q

Challis

Tr

Custer graben

Mt. Greylock 9857 ▲

Yankee Fork

Bald Mtn. 10,313 ▲

Germer Peak 7684 ▲

Warm Spring Creek graben

M

93

Sawtooth fault

Basin Butte 8854 ▲

21

Sawtooth Valley

Sunbeam

Salmon River

Clayton

Lone Pine Peak 9655 ▲

Tr

Stanley

Lookout Mtn. 9954 ▲

White Cloud Peaks

M

Redfish Lake

Tg

Q

Obsidian

▲ 9847 Decker Peak

Parks Peak 10,208 ▲

Castle Peak ▲ 11,815

Challis volcanic rocks

Tr

Q

tight folding in white Ordovician quartzite exposed in a big cliff above road

Sawtooth Range

Sawtooth batholith

75

Snowyside Peak ▲ 10,651

Alturas Lake

U Pal

Galena Peak ▲ 11,170

Galena Summit

Ryan Pk ▲ 11,795

Tr

Idaho Atlanta batholith

Norton Peak 10,285 ▲

Boulder Peak ▲ 10,906

Boulder Mountains

Pioneer Mtn. Stock

pink 44 million-year-old granite of the Sawtooth batholith contrasts with older gray granite of the Idaho batholith surrounding it

Smoky Mountains

Big Pk. ▲ 10,060

Tr

Bald Mtn. ▲ 9280

Sun Valley

Ketchum

Hyndman Peak ▲ 12,078

Soldier Mtns batholith

U Pal

Kelly Mtn. 8827 ▲

Hailey

Bellevue

▲ 7565 Lookout Mtn.

Tr

M

gently sloping former desert surface stands above highway on east; light gray limestone cliffs in mountains above

Q

Moonstone Mtn. 6144 ▲

Q

20

93

Clarendon Hot Springs stock

Magic Reservoir

Magic Reservoir caldera erupted rhyolite ash-flow tuffs and collapsed in Pliocene time, about 3 to 6 million years ago

Plioc rhy

Shoshone Ice Caves

recent basalt flows

Qb

Idaho 75
Shoshone—Challis

75

recent basalt flows

Snake River Plain

Wedge Butte rhyolite is about 3 million years old

Shoshone

93

10   0   10   20 MILES

10   0   10   20   30 KILOMETERS

224

between 70 and 90 million years ago. Then, about 50 million years ago, more granite intruded the region and great piles of Challis volcanic rocks blanketed much of the older bedrock.

The most conspicuous rocks along the road between Bellevue and the southern end of the Stanley Basin are Paleozoic sedimentary formations, which tend to be very dark because they contain organic matter. That same stretch of road also passes volcanic rocks, mostly rhyolite, which tends to be pale. Watch the colors. Small areas of granite, another pale rock, betray themselves by their bouldery outcrops.

Deep deposits of sediment, much of it glacial outwash, floor most of the Stanley Basin. Rocks in the mountains on both sides of the basin and into the area a few miles east of Sunbeam are all granite, most of it the Idaho batholith. Large masses of the younger granite emplaced about 50 million years ago exist in the Sawtooth Range west and south of Stanley.

Older rocks between the area a few miles east of Sunbeam and Challis are folded Paleozoic formations, now largely buried under pale Challis volcanic rocks erupted about 50 million years ago.

Metamorphic Rocks in the Pioneer Mountains

Hyndman Peak is the outstanding landmark in the high Pioneer Mountains northeast of Ketchum. Those lofty mountains contain a large mass of thoroughly metamorphosed sedimentary rocks. Scattered age dates suggest that they were cooked during Paleozoic time. If more age dates confirm that suggestion, it will become reasonable to associate that metamorphism with the Antler orogeny. That enigmatic event of middle Paleozoic time left its vague footprints here and there in Idaho, everywhere trampled almost beyond recognition by later events.

During Eocene time, about 50 million years ago, masses of granite magma invaded the much older metamorphic rocks. Then, sometime after Eocene time, something stretched the earth's crust in this region, pulling the cover of younger rocks off the metamorphic rocks and granite beneath. A zone of strongly sheared rock that nearly encircles the metamorphic rocks appears to be the fault along which the rocks moved. Little structural details in the rocks show that the rocks that formerly covered the metamorphic rocks moved northwest.

Granite in the Summit Creek stock looks exactly like that in the mass that intruded the metamorphic rocks 12 miles to the southwest. Geologists familiar with both granites suggest that they are in fact the

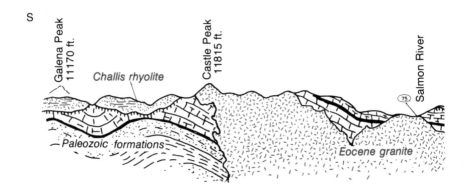

Section drawn across the line of the

same, that the Summit Creek stock is the beheaded top of the granite intrusion in the metamorphic rocks. If so, fault movement within the sheared zone amounts to about 12 miles.

Challis Volcanic Rocks

The Challis volcanic rocks, mostly rhyolite, erupted across a broad area of central Idaho around 50 million years ago. Several thousand feet of the pale ash still blanket much of the region. The road passes excellent exposures of that ash in the area between Clayton and Challis. Watch for white and yellowish outcrops of rather soft rocks that show up best in stream banks and road cuts.

Like most large rhyolite volcanoes, these were almost unbelievably violent. It is hard to exaggerate the explosive potential of large volumes of red hot steam. The Challis eruptions continued for a long time, and finally produced a complex pattern of more or less overlapping calderas that probably did not appear as distinctive features in the landscape even when they were new. Now, after some 50 million years of erosion, it is impossible to point to a mountain and identify it as an old volcano. Instead, you see an ordinary erosional landscape carved into volcanic rocks.

Stanley Basin and the Sawtooth Range

The steep front of the Sawtooth Range, the scarp of the Sawtooth fault, forms a straight wall that defines the west side of the Stanley

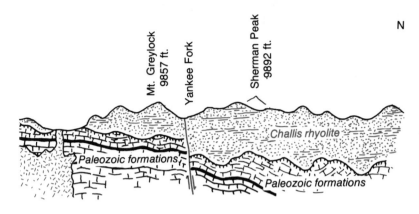

road between Sunbeam and Clayton.

Basin. Movement along the Sawtooth fault raised the crustal block of the Sawtooth Range while simultaneously dropping the Stanley Basin.

A few of the outcrops in the mountain front facing the Stanley Basin expose a complex of metamorphic and igneous rocks that probably belong to the extremely ancient Precambrian basement. Most of the rocks west of the mountain front are granite. The pale gray granite of the Idaho batholith was emplaced 70 million years ago, the pink granite of the Sawtooth batholith about 20 million years later. Similar rocks probably lie beneath the deep gravel deposits in the floor of the Stanley Basin.

The glaciated front of the Sawtooth Range near Stanley.

227

Pink pegmatite cutting light gray granite east of Stanley.

The Wood River Mines

Ketchum started its career in 1880 as a milling and smelting town named Leadville, the vital center of the Warm Springs mining district. Ore was discovered in the Wood River area considerably earlier, sometime during the 1860s, but nothing much happened until about 1880, when the district launched a decade of major production.

Most of the mines cluster in three areas: The Hailey-Bellevue mineral belt in the hills a mile or two south of Hailey, the Triumph-Parker mineral belt just east of Ketchum, and the Mayflower-Red Elephant area about two miles west of Hailey. All were active mainly during the 1880s, sporadically since then. The Triumph Mine, northeast of Hailey, came late. It produced more ore than all the others combined between 1927 and about 1950.

The district began to produce barite in 1947 from the Sun Valley Mine near Hailey. That ore is in veins in the same sedimentary rocks that earlier produced lead and silver. Barite is a heavy mineral used mostly as an additive in drilling mud for oil wells. The great density of the barite makes the mud heavy so that it can keep the lid on high pressure gas. Barite mud prevents gushers.

The Wood River District produced lead and silver ores from the Milligen and Dollarhide formations, several thousand feet of slaty black shale and limestone. The original sediments of the Milligen formation, mostly east of the road, were deposited in shallow sea water during Mississippian time, approximately 340 million years ago. The Dollarhide formation, which is mostly west of the road, accumulated during Permian time, about 260 million years ago.

Then, as the Rocky Mountains formed during late Cretaceous time, a number of igneous intrusions invaded the area, bringing heat that baked and somewhat recrystallized the older rocks. That slight metamorphism converted the organic matter in the rocks into graphite, which colors the rock so intensely black that some of the mine dumps look like piles of coal. Meanwhile, mineralized solutions circulating through and around the granite intrusions added ore minerals.

Gold in the Yankee Fork District

Prospectors found placer gold in the Yankee Fork, north of Sunbeam, in 1870. Mining began in 1873. The town of Bonanza sprang up in 1877, Custer a year or so later. Then a gold rush in 1879 raised the population of Bonanza to more than 2000. The best gravel had been washed by the turn of the century, and Bonanza and Custer were ghost towns a few years later.

The largest dredge ever to operate in Idaho converted more than five miles of the Yankee Fork into long windrows of raw gravel between 1948 and 1952. The total take from that operation is officially reported to have been a little more than one million dollars. It is hard to imagine how that could have included much profit, considering the cost of the dredge and the expense of operating it. Also consider the long term value of the five miles of stream and floodplain it devastated, all the fish that no one will ever catch, all the trees and grass that will never grow. The monster dredge still rests in its last pit about a half mile north of the old town of Bonanza, eight miles north of the highway. It was donated to the state of Idaho in 1966 for use as a mining museum.

Properly managed gold dredges can be fiendishly efficient placer mining machines — there is some question whether the one that worked the Yankee Fork was properly managed. Gold dredges use a chain of steel buckets to scoop gravel out of the floodplain and dump it into sluice boxes, where a stream of water washes it over a series of riffles. The heavy gold settles behind the riffles while the pebbles and sand continue on their way. Some dredge captains kept a little pool of mercury behind each riffle to amalgamate the gold, making doubly sure that small flecks stayed put. Every few days, the captain stopped mining for a few hours while he cleaned the nuggets and amalgam from behind the riffles. After the gravel left the sluice box, it landed on a conveyor belt that dumped it off the after end of some dredges, off to either side of others. Watch for the little cross ridges on the gravel ridges that record the swings the conveyor belt made as it stacked the gravel tailings.

Most gold dredges could dig to a depth of about 35 feet, enough to reach bedrock in most small streams. It is essential to mine the deeper gravels and scrape the bedrock beneath them, because that is where most of the gold lies. Irregularities in the bedrock surface function as natural riffles. Before gold dredges, placer miners mined underground to reach bedrock, a desperately dangerous expedient that requires a great deal of expensive timbering and pumping.

So the gold dredge, which could float in about ten feet of water, navigated through the floodplain by digging forward and filling behind until it finally ran into hard bedrock or out of rich gravel. Dredge crews came to consider themselves quite the seamen, and used all sorts of nautical titles and terms that sound far more appropriate to a deepwater ship than to a gold mine.

Early placer miners in Jordan Creek, which flows into the Yankee Fork, opened underground workings during the 1870s to reach the deeper gravels and the bedrock surface beneath them. Their tunnelling through the gravel in Jordan Creek revealed some of the richest placers and biggest nuggets ever found in Idaho. One especially handsome specimen weighed two pounds. A dragline worked a little more than a mile of the floodplain of Jordan Creek between 1948 and 1950. That machine is officially reported to have recovered about a quarter of a million dollars worth of gold, but there are those who darkly suggest that some of the nuggets may have been diverted to unofficial destinations.

Prospectors looking for the "mother lode" in the Yankee Fork area found their first bedrock vein in 1875, another a year later, both in Challis volcanic rocks. Both veins were well exposed, so miners could begin working them with a minimum of preliminary development. In fact, erosion had stripped the covering rock off most of the General Custer vein, leaving it exposed on the slope of a hill. Some of the early ore proved rich enough that the operators could afford to bag it and ship it to Salt Lake City for treatment. Shipping costs were almost unbelievably high in those days of wagon trains, so that must have been real bonanza rock. Arrastras were used in the early days to grind ore locally, but those mule powered contraptions never could process much rock. Completion in 1881 of a big stamp mill enormously increased production.

The glory days ended in the Yankee Fork district about 1905, and most of the people had gone by 1910. Estimates place total production at approximately 12 million old fashioned dollars worth of gold and silver. Mining has been sporadic and disappointing since 1905, despite periodic excitements and occasional new mines and mills.

The Bayhorse District

Prospectors discovered ore in the Bayhorse district, in the hills about a dozen miles north of Clayton, in 1872, and mining began a few years later. According to most estimates, the district produced more than 10 million dollars worth of lead, zinc, and silver ore. That makes it Idaho's second largest producer, after the Coeur d'Alene district, of all three metals — a poor second.

Bayhorse ore was in veins in Paleozoic sedimentary formations. Most had been enriched by weathering: metals dissolved near the surface moved down the vein in trickling ground water, then redeposited just below the water table, a process called supergene enrichment. Ore bodies of that type tend to be extremely rich, but they never persist in depth because the redeposited metals concentrate in a very narrow zone. So the mines that work them typically enjoy brief flings of thrilling prosperity as they work through the high grade ore, then the sudden chill of poverty as they get into the much leaner ore of the original vein.

The grassy hills in front of the Sawtooth Range are glacial moraines. View west of the highway a few miles south of Stanley.

Although the lavish years ended in 1898, the Bayhorse district revived several times: in 1910, from 1920 to 1925, in 1935, and again during the late 1960s. During the late 1940s mines in the northern part of the Bayhorse district produced a few hundred tons of fluorite.

The Thompson Creek Mine

The Thompson Creek Mine is about five miles northwest of the Yankee Fork Ranger Station. The mine was developed during the 1970s as an open pit that was to produce some 25,000 tons of molybdenum ore a day from a deposit that contains at least 185 million tons of ore. It would have been the largest mine in Idaho, the source of about one-tenth of the world's supply of molybdenum.

Unfortunately, the price of molybdenum dropped so drastically while the deposit was under development that the mine never went into full production. The ore is still there, and someday, when the price of molybdenum rises, the mine will bring another boom to central Idaho.

Uranium

Small quantities of uranium ore, most of it fairly low grade, exist in quite a number of places in the hills northeast of Stanley. The uranium minerals occur in veins that fill fractures in granite of the Idaho batholith, the younger Challis volcanic rocks, and in sandstone between them. Although several small mines did ship some ore, none produced much.

Glaciation

The serrated ridgeline and deeply gouged valleys of the high Sawtooth Range west of the Stanley Basin tell of severe mountain glaciation. Glaciers came down to and slightly beyond the mouths of the valleys, where they left their great moraines that now form the forested hills along the base of the range. The ice did not advance onto the floor of the Stanley Basin. Stanley, Redfish, Pettit, and Alturas lakes all nestle within moraines along the fault scarp at the front of the range.

Mountains east of Stanley Basin are as high as the Sawtooth Range, but glaciated only in their highest peaks. Presumably the peaks of the Sawtooth Range snatched most of the snow out of the clouds before they got that far east. Meltwater from the glaciers in both ranges spread deep deposits of gravel across large parts of Stanley Basin, and down the Salmon River.

As elsewhere in the northern Rocky Mountains, there are two sets of moraines: a larger set that records the earlier glaciation, and a smaller set left from the most recent ice age. Widely scattered deposits of very deeply weathered glacial till record at least one much earlier ice age of which almost nothing is known.

Glacially sculptured peaks of the Sawtooth Range reflected in Stanley Lake. — U.S. Geological Survey photo by Charles Miles

233

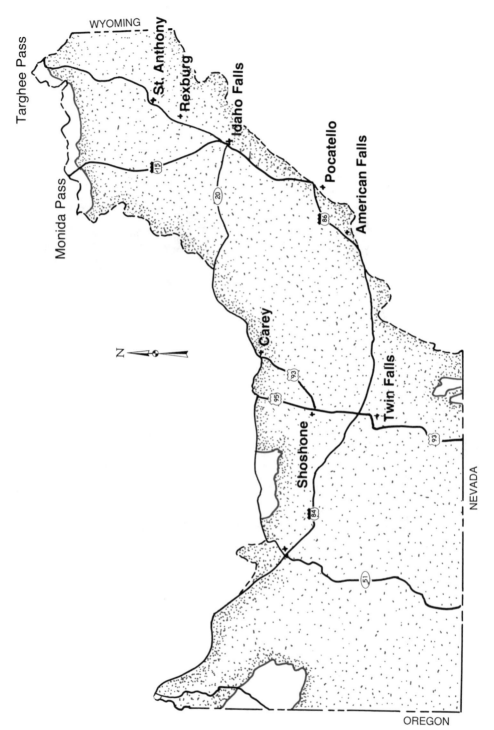

Roads covered in this section.

IV:
The Snake River Plain

The Snake River Plain looks on the geologic map like a broad stroke of a brush that starts in southwestern Idaho and narrows as it swings northeast to a sharp point in the Yellowstone volcano. On the ground, it looks at first glance like a broad and nearly featureless prairie, a dreary vastness of flat sagebrush desert somehow caught between distant mountains. Neither first impression is accurate; the Snake River Plain is far more varied and interesting than either might suggest. In fact, it is a major geologic spectacular, one of the world's largest and most intriguing volcanic provinces.

Because most of the exposed bedrock in its central and eastern parts is basalt lava flows, early geologists called the Snake River Plain a basalt plateau, envisioning it as something similar to the Columbia Plateau. That view was, to put it mildly, wrong.

Years of more careful observation have now shown that eruptions in the Snake River Plain actually produced enormous volumes of rhyolite, modest quantities of basalt. Furthermore, the basalt eruptions did not produce overwhelming floods of lava like those that built the Columbia Plateau. The perfectly normal basalt flows in the Snake River Plain barely cover the much larger volumes of rhyolite beneath. The province consists primarily of rhyolite rather well hidden under a thin facade of basalt. Despite its mostly basalt surface, the Snake River Plain does not resemble the Columbia Plateau. It is utterly different.

The thin veneer of black basalt that caps a thick section of rhyolite in Jackass Butte in Owyhee County is typical of the Snake River Plain.
— U.S. Geological Survey photo by H.E. Malde

Neither is the Snake River Plain quite the smooth sweep of a brush that a casual glance at the geologic map suggests. Think of it instead as two provinces: the main trend of the Snake River Plain that follows a nearly straight line from near the southwestern corner of Idaho to Yellowstone Park, and the Western Snake River Plain that reaches northwest from the main trend. They differ more than the view from the road might lead you to suspect.

The Western Snake River Plain

The western part of the Snake River Plain trends slightly west of north, parallel to the trend of Basin and Range valleys. In fact, it is one of those valleys, a Basin and Range fault block that filled with white rhyolite ash, black basalt lava flows, and a rich assortment of valley-fill sediments as it dropped between faults.

Travelers crossing the western Snake River Plain see remarkably little bedrock. The valley-fill sediments and much of the rhyolite ash are still not well enough hardened into rock to

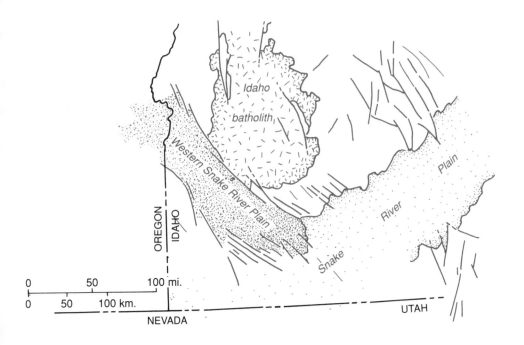

The western Snake River Plain is a northern projection quite separate from the rest of the province. It is a Basin and Range valley.

make resistant outcrops, so you see them only in occasional road cuts and stream banks. Here and there, more resistant basalt lava flows set flat caps on low hills.

The Main Snake River Plain — the Track of a Migrating Hotspot

The main trend of the Snake River Plain consists almost entirely of rhyolite erupted from a long row of extinct volcanoes that started in southwestern Idaho about 13 million years ago. The volcanoes become progressively younger northeastward along a nearly straight line that ends in the Yellowstone volcano, which is active.

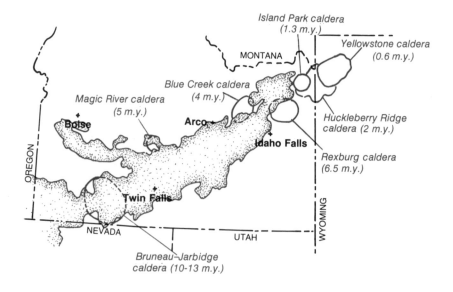

The trend of the volcanic centers that built most of the Snake River Plain.

That long volcanic track has led many geologists to regard the Yellowstone volcano as the leading prong of the Snake River Plain. They propose a most appealing model in which a series of giant rhyolite volcanoes essentially like the one now active in Yellowstone Park formed, erupted for a million or two years, then became extinct as a new volcano began to erupt a few miles farther northeast. That steady northeastward march of the volcanic center evidently records the steady southwestward movement of the North American continent over a stationary hotspot in the mantle. The volcanic center migrates northeast as the North American plate moves southwest just as a saw cut moves from left to right through a board as you shove it into the saw from right to left.

We think the hotspot formed about 17 million years ago, in late Miocene time, when a large meteorite struck southeastern Oregon; exploded with enough violence to open a crater that bit deeply into the upper mantle. Opening the crater relieved pressure on the already hot rocks in the upper mantle, permitting them to begin to melt to form basalt magma. The crater flooded with molten basalt to become a lava lake, which overflowed into northern Oregon and Washington in the enormous flood basalt flows that built the northern part of the Columbia Plateau.

As the eruptions partially emptied the lava lake, they continued to relieve pressure on the rocks at depth, which then continued to melt to produce more basalt magma. Rock constantly rises from deep within the mantle to replace the erupting magma, so the eruptions maintain a rising column of hot rock within the mantle. That process continues even though continental crust now covers the mantle site of the crater. The wound in the mantle still festers with basalt magma.

To put it simply, the system continues to erupt because it continues to erupt. Each eruption relieves pressure on the hot rocks in the upper mantle, thus permitting them to partially melt to form more magma. Meanwhile, hot rock rises from deeper in the mantle to replace that lost to eruptions. The character of the eruptions changed about 13 million years ago, when plate movement finally brought uncratered continental crust over the rising column of hot rock in the mantle. Now that hot rock rising from deep within the mantle is melting the passing continental crust as though it were a sheet of boiler plate slowly passing over a welding torch.

If the Snake River Plain is the volcanic track left as the continent moves across a hot spot in the mantle, then the eastward displacement of the volcanic center should match the westward movement of the continent. It does. The Yellowstone volcano is about 300 miles northeast of southwestern Idaho, so the volcanic center has been moving at a rate of about an inch and a half per year. That closely fits the rate at which most geologists estimate the North American plate is moving west from the mid-Atlantic ridge, about two inches per year. The numbers match as well as anyone could wish in the imperfect business of geologic measurement.

The Yellowstone Volcano

The Yellowstone volcano is now directly over the hotspot, the mantle site of the meteorite impact. It is one of the largest and most violent active volcanoes in the world, one of a small number of those rhyolite monsters known as resurgent calderas. They are, by any standard, the most outrageous of volcanoes, perfect horrors.

Resurgent calderas typically erupt enormous volumes of rhyolite lava in each of several enormous explosions spaced at intervals of hundreds of thousands of years. The amount of lava they produce during each heroic outburst ranges from a few dozen to as many as 200 cubic miles. The ground surface subsides as all that rhyolite erupts to open an enormous collapse basin called a caldera that may be as much as several tens of miles across. Then, continuing volcanic activity fills the caldera with rhyolite, along with minor amounts of basalt. The next big series of eruptions will open and then largely fill a new caldera.

The three enormous calderas that formed in the last 2 million years.

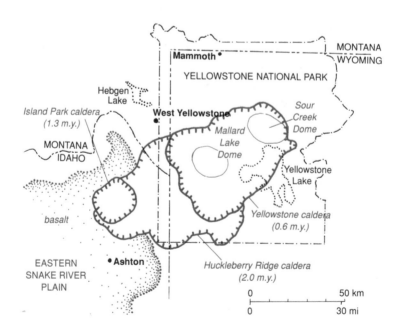

240

So far, the Yellowstone resurgent caldera has gone into major eruption three times at intervals of approximately 600,000 years, most recently about 600,000 years ago. That long lapse since the last major eruption gives little reason to assume that the volcano is finished, little cause for comfort. Quite the contrary. A large mass of molten magma, almost certainly rhyolite, now exists a few thousand feet beneath the park. All the ingredients for a new eruption are in place.

Two large areas within Yellowstone Park are now bulging like slowly growing tumors, evidently because magma is rising beneath them. Swarms of small earthquakes rattle the park from time to time, and some of the thermal areas are growing larger and hotter. Those are all signs of continuing volcanic activity, typical symptoms of impending eruption; they provide excellent reason to expect the future to bring more eruptions.

Although rhyolite ash looks harmless enough in its pastel shades of pale gray, yellow, and pink, the eruptions that produce it are extremely violent, extraordinarily dangerous. Molten rhyolite magma combines the volcanically perilous properties of high viscosity and a burning thirst for water.

In fact, rhyolite magma is so viscous that it has a consistency almost like that of modelling clay; it hardly seems believably liquid. That extraordinary viscosity would be perfectly harmless if rhyolite magma did not also tend to absorb large quantities of water — at magmatic temperatures that water translates into red hot steam. Steam can hardly bubble quietly out of such viscous magma to blow off in great flames licking from the throat of the volcano. Instead, steaming rhyolite magma expands internally as it erupts, like rising bread, but so rapidly that it explodes like a grossly oversized and overheated steam boiler.

Glowing clouds of finely shredded rhyolite lava suspended in red hot steam boil out of the devil's cauldron of a collapsing caldera, suddenly pouring molten volcanic ash across hundreds or thousands of square miles of surrounding countryside. That is an ash flow. Most ash flows are still so hot when they finally quit moving that the shreds of lava weld themselves together into a solid rock. Meanwhile, blasts of steam also blow ash high into the atmosphere, where it drifts downwind and eventually settles to cover the countryside like snow. That is an ash fall.

No resurgent caldera has erupted during the period of written history; we have no eyewitness accounts that might tell us what to expect. In fact, there would probably be some considerable difficulty in finding surviving eyewitnesses if one of those monsters were to erupt. To get a rough idea, consider that the 1980 explosion of Mount St. Helens produced less than one-third of a cubic mile of rhyolite. Compare that to the 200 or more cubic miles known to have blown from some resurgent calderas during a single eruption. If the Yellowstone volcano does erupt again, it is likely to cause an absolutely appalling natural disaster, one far greater than any in historically recorded human experience.

The Volcanic Lowland

The broad swath of the Snake River Plain is distinctly lower than the mountains that border it on the north and south. We think that furrow through the mountains reflects the extreme violence of the rhyolite eruptions.

Throughout most of the High Plains east of the Rocky Mountains geologists find thick beds of rhyolite ash, many of which they call the Pearlette ash. It is the same age as the Snake River Plain. We have never seen a careful estimate of the total volume of Pearlette ash; our own extremely approximate estimate suggests that if it were all swept up and dumped on the Snake River Plain, it would fill the furrow about to the level of the mountains that surround the Snake River Plain. We suggest that the Snake River Plain is low because so much of the continental crust that melted above the Yellowstone hotspot exploded into the atmosphere, then drifted downwind to land on the High Plains as air-fall ash.

Second Generation Volcanoes:
The Basin and Range

If the Snake River Plain consists mostly of rhyolite, then why is basalt the most visible rock? Why does that thin veneer of basalt lava flows cover the rhyolite almost everywhere? The most recent basalt flows erupted along the Great Rift, a line of

The Snake River Plain in miniature: A road cut reveals black basalt lying on pale rhyolite, the handiwork of the Basin and Range lying on that of the Yellowstone hotspot.

open fissures that extends from Craters of the Moon National Monument southeast almost to American Falls. Why does volcanic activity continue in a part of the Snake River Plain that passed over the Yellowstone hotspot several million years ago? Evidently, the hotspot theory doesn't tell the whole story of Idaho's more recent volcanic activity.

Another way to produce large volumes of magma is to stretch the earth's crust, thus reducing pressure on the rocks in the upper mantle, which are already so hot that they would melt if they were not under such great pressure. The open cracks of the

Some of the youngest basalt lava fields on the Snake River Plain.

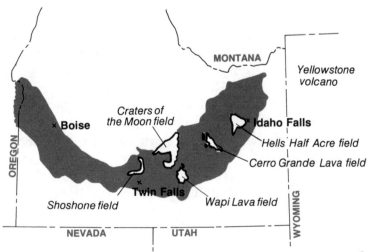

243

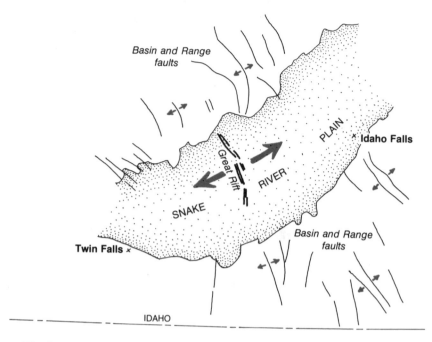

The Great Rift. The arrows indicate the direction of crustal stretching that opened it.

Great Rift leave no doubt that the earth's crust in the Snake River Plain is stretching. That certainly reduces the pressure on the rocks at depth, almost certainly causes the rocks in the upper mantle to begin melting to produce basalt magma. If the cracks explain the magma, what explains the cracks?

The Great Rift trends parallel to the faults that define the mountain and valley blocks of the Basin and Range both north and south of the Snake River Plain. Evidently, the Basin and Range faults are just beginning to dismember the Snake River Plain into mountain and valley blocks. The entire Basin and Range is a province of both crustal stretching, and of widespread volcanic activity. We suggest that the later basalt eruptions in the Snake River Plain fit into the picture of the Basin and Range, that the volcanoes along the Great Rift are comparable to those scattered here and there throughout the Basin and Range, not the direct progeny of the hotspot that created the Snake River Plain. But that view raises the obvious question of why basalt is so much more abundant in the Snake River Plain than elsewhere in the Basin and Range.

244

We believe that the deeper levels of the Snake River Plain probably consist of granite, that millions of years of erosion will eventually strip the volcanic rocks off its surface to reveal a long track of granite. We imagine that the Yellowstone hotspot may be leaving a broad smear of overheated mantle under that granite. When Basin and Range faults begin to break the Snake River Plain that already hot mantle melts very easily, and then erupts basalt lava.

Basalt Flows

Basalt lava is about fluid enough to run like molasses, and the stuff rarely contains all that much steam. It generally spreads out into relatively thin flows that cover a large area with a minimum of commotion. Most basalt flows are between 10 and 50 feet thick, and they quite commonly cover an area of several square miles. That may sound like plenty of lava, but it is merely a drop compared to the flood basalt flows that covered thousands or tens of thousands of square miles in the Columbia Plateau.

As basalt flows begin to crystallize, they go through an interesting stage in which the top of the flow is fairly solid while the interior is still fluid. The moving lava within quite commonly buckles the solidifying surface into broad pressure ridges — look along their crests for places where oozes of molten basalt squeezed up through cracks. Many pressure ridges appear to form in places where the flow moves across irregularities in the ground surface.

A pressure ridge in basalt, beside U.S. 93, near Carey.

Molten lava may also drain out from under the solid crust, leaving the flow hollow. Lava tubes that may be as much as several miles long form within the flow. They seem like tunnels when you walk through them. Those caves eventually collapse to form rows of sinkholes in the surface of the flow.

If basalt lava absorbs a bit of water before it erupts, the steam escapes quite easily from the fluid magma, blowing a dark plume of ash and volcanic cinders out of the vent to build a small volcano, a cinder cone. Although those eruptions make wonderful natural fireworks displays, they are not actually violent — by volcanic standards. Basalt magma never explodes violently.

Over a period that may last from a few days to as long as several months, the cinder cone grows to a height that rarely exceeds a few hundred feet. The aerial fireworks stop as the erupting magma blows off the last of its head of steam. Then the remaining magma generally erupts as a lava flow, perhaps as several flows.

Flows typically burst from the bases of cinder cones, instead of from the crater in the summit. That happens because the loose pile of cinders is too weak to contain the rising column of basalt. Large sections of the cinder cone commonly raft away on the moving flow, leaving behind a partially dismantled wreck of a volcano that may be very hard to recognize.

Cinder cones are typically single shot volcanoes that rarely erupt a second time. In all likelihood, a quiet cinder cone is truly dead. Further eruptions in the same area ordinarily build new cinder cones, each with its own flows. The sudden appearance of a new cinder cone volcano should not come as too much of a shock in an area that already contains many younger cinder cones.

If a series of basalt eruptions does issue from the same vent, it builds a very broad and gently sloping pile, a shield volcano. The eastern Snake River Plain is a broad field of small shield volcanoes, dozens of them. Most make such subtle rises in the plain that you hardly notice them until you know they exist. Then you see them all around. Many have small cinder cones on their crests, relics of the latest eruption.

That Nice Fresh Volcanic Surface

Except for the deep canyons of the Snake River and a few of its tributaries, most of the Snake River Plain has nearly escaped erosion. Much of the original volcanic surface remains nearly intact. The volcanic rocks are too young to have eroded very much, and the internal structure of basalt lava flows makes them difficult to erode. The conspicuous tendency of basalt flows to break into vertical columns opens passageways that encourage surface water to sink into the flow without lingering long enough to weather the rock into clay. Furthermore, many flows in the Snake River Plain contain lava tubes, which absorb great volumes of water, thus further protecting the surface from erosion.

Basalt columns near Buhl. — U.S. Geological Survey photo by H.E. Malde

The Snake River Plain owes its dryness almost as much to the thirstiness of basalt lava flows as to the meager rainfall. In fact, the flows even absorb the streams that drain onto them from the mountains of central Idaho, making it impossible for them to carve valleys across the Snake River Plain.

Once underground, the water soaks into the rhyolite ash and collects in pore spaces in the zones of buried soil and broken rubble between lava flows. Even where the surface of the Snake River Plain is extremely dry, it is commonly possible to get a good well. The stored ground water slowly migrates down the gentle southward slope of the volcanic rocks, eventually to seep out through the uncountable springs along the north wall of the Snake River Canyon.

Weathering of the basalt will eventually produce enough clay to seal the flows so water can no longer soak into them. Only then will some of the rain begin to run off as surface water, and start to erode valleys. If the older Columbia Plateau, which is just starting to erode, is any guide, that will take another several million years, at least.

THE ICE AGES

Although the ice ages came and went without spreading glaciers across any part of the Snake River Plain, they did leave their mark on the landscape. The wetter ice age climates filled the Salt Lake basin with enough water to overflow into the Snake River in one of the greatest floods of known geologic record. And the colder ice age climates left their mark in large expanses of patterned ground similar to that now active in the Arctic.

The Great Bonneville Flood

Evaporation is the only outlet for the Great Salt Lake, the puddle that now floods the lowest part of the Salt Lake basin. Ice age climates were wet enough to fill that basin to form Lake Bonneville, an inland sea fully comparable to the modern Great Lakes. The water rose until it finally overflowed through Red

Red Rock Pass as it appeared in an old woodcut that the Geological Survey published a century ago. Rapid erosion of the narrow gorge released the Bonneville flood.

Rock Pass, and into the Snake River drainage. The Bonneville shoreline, at an elevation of 5085 feet, records the lake as it was until overflow through the pass cut through a hard layer of rock into much softer material below. Rapid erosion in the pass then released the Bonneville flood, which continued until a lower mass of resistant rock stabilized the lake at the less conspicuous Provo shoreline, about 300 feet lower. Evidence of that catastrophic drainage appears along the entire length of the Snake River canyon.

The muddy flood filled the entire canyon of the Snake River above its brim, and grumbled across the neighboring surface of the Snake River Plain. It raged through the narrow parts of the canyon, and ponded more or less quietly in the broad valleys between narrows. The flood scoured loose rock out of the entire length of the canyon, and scrubbed the canyon walls, while ripping their basalt columns apart, chunk by chunk. It dumped most of the debris where the flow slackened as it entered the upper parts of the broad valleys.

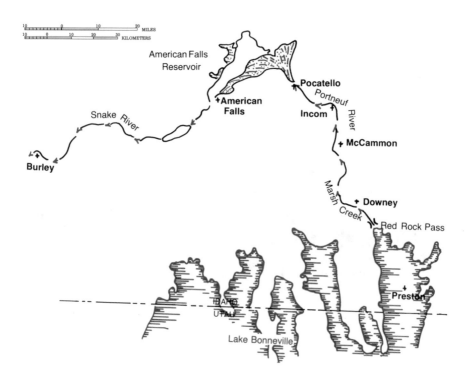

Lake Bonneville drained through Red Rock Pass into the Snake River.

The most spectacular Bonneville flood deposits are giant gravel bars a mile or more long and more than 100 feet high that look at first glance like hills within the canyon. A closer look reveals that they are greatly magnified versions of the gravel bars you see in a normal stream. The giant bars generally lie near the heads of the broad valleys, and have a tapering form streamlined in the direction of water flow. Rounded boulders of black basalt far larger than anything the modern Snake River can move litter their surfaces like petrified crops of black watermelons.

The best fields of watermelon boulders are within a mile downstream from one of the narrows, where the flood tore the pieces of basalt off the canyon walls. Melon boulders abound more on the surfaces of the bars than within because the waning flood continued to carry finer sediments long after it dropped the big stuff. The melon boulders lagged behind.

250

Narrower reaches of the Snake River Canyon where the water ran fastest look freshly scrubbed; they generally lack the accumulations of loose talus that line the walls of most narrow canyons in Idaho. Giant whirlpools plucked deep holes in the canyon floor; the Snake River now pours over waterfalls into many of those pools. Big deposits of fresh gravel that cover long stretches of the surface on both sides of the canyon show that the flood overflowed onto the flat ground beside it.

The simplest way to measure the amount of water going down a stream is to multiply the speed of stream flow by the area of the channel. It is easy enough to measure the area of the channel, especially in narrow parts of the canyon where the upper limit of flood erosion is clearly marked. You can estimate the flow speed of an anciently defunct stream by measuring the largest rocks it moved, in this case the rounded melon rocks. Remember though that any such estimate gives a minimum result because the size of the rocks depends as much upon the spacing of fractures in the bedrock as upon the speed of stream flow. Unless you find loose chunks of rock that were too big to move, you have not seen the upper limit of what the flood was capable of moving. So far as we know, the Bonneville flood moved every loose rock in its path.

Boulders of the melon gravel litter the surface of a giant bar in the Snake River Canyon. — U.S. Geological Survey photo by H.E. Malde

The best informed estimates have the Bonneville flood at its peak discharging water down the Snake River Canyon at a rate something like one-third of a cubic mile per hour. That is approximately three times the discharge of the Amazon River. To our knowledge, the only greater flood flows were those released when Glacial Lake Missoula suddenly drained. Those probably discharged between eight and ten cubic miles of water per hour.

The total volume of water was approximately 600 cubic miles, comparable to one of the smaller Great Lakes, and to Glacial Lake Missoula. All that water went down the Snake River within a few months, at most. Had you been there to watch the flood pass, you could have listened to the sound of melon boulders bouncing along the canyon floor in a chorus of dull booms.

Patterned Ground

Water expands as it freezes into ice, then the ice contracts as the temperature drops below freezing. So frozen ground shrinks in very cold weather. As it shrinks, the ground breaks into a polygonal pattern of fractures like those in sun cracked mud, except that the soil polygons are the size of rooms, or larger. In very cold regions where the subsoil remains permanently frozen, those fractures open during the coldest months of every winter, letting more rocks fall into them. Finally, the fractures

Patterned ground in Elmore County. The sagebrush and the hammer just left of center indicate the scale. — U.S. Geological Survey photo by H.E. Malde

develop into a polygonal pattern of stony stripes on the ground surface, patterned ground. That happened in large parts of the Snake River Plain during the ice ages.

The ground is no longer permanently frozen in the Snake River Plain, and soil polygons no longer develop there, but old soil polygons still cast their network pattern here and there. Watch for those souvenirs of the ice ages in areas that have never been plowed. Chunks of black basalt show through the yellow grass in lines and irregular polygonal patterns that vaguely suggest a giant net cast over the ground.

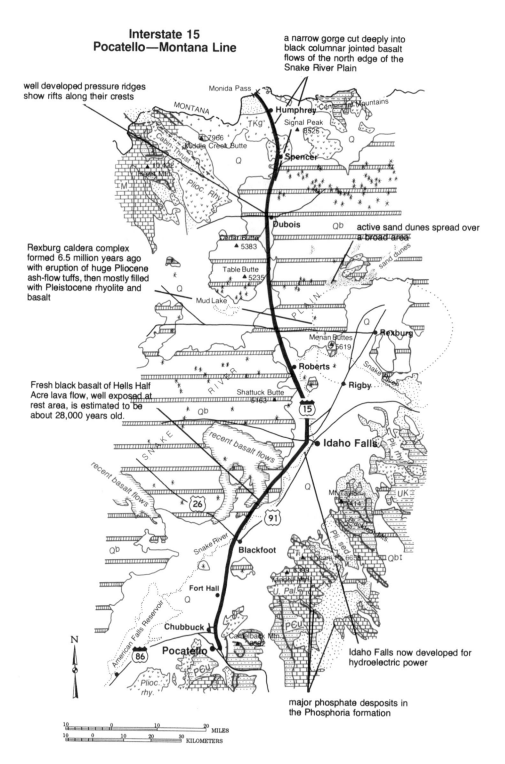

Interstate 15
Pocatello—Montana Line

a narrow gorge cut deeply into black columnar jointed basalt flows of the north edge of the Snake River Plain

well developed pressure ridges show rifts along their crests

Monida Pass

MONTANA

TKg

Humphrey

Centennial Mountains

Signal Peak
▲ 8526

Cabin Thrust

▲ 7966

Middle Creek Butte

Q

Spencer

Plioc. Rhy.

▲ 10 132
Italian Mtn

TMJ

Dubois

Qb

active sand dunes spread over a broad area

Rexburg caldera complex formed 6.5 million years ago with eruption of huge Pliocene ash-flow tuffs, then mostly filled with Pleistocene rhyolite and basalt

Cedar Butte
▲ 5383

sand dunes

Table Butte
▲ 5235

Q

Mud Lake

P L A I N

Q

Rexburg

Menan Buttes
▲ 5619

Roberts

Snake River

Fresh black basalt of Hells Half Acre lava flow, well exposed at rest area, is estimated to be about 28,000 years old.

Shattuck Butte
▲ 5163

Rigby

S N A K E

15

recent basalt flows

Qb

Idaho Falls

recent basalt flows

R I V E R

Pil. Mtn

Mt Taylor
▲ 414

UK

26

91

Pil. sed

Qbl

Qb

Snake River

Blackfoot

Big Southern

Hibbard Mtn 6659

Yandell Mtn

U. Pal

Fort Hall

Q

PCu

American Falls Reservoir

Chubbuck

Cannelback Mtn
▲ 6652

N

86

Pocatello

Idaho Falls now developed for hydroelectric power

PCu

Plioc. rhy.

major phosphate deposits in the Phosphoria formation

10   0   10   20 MILES
10   0   10   20   30 KILOMETERS

Interstate 15:
Pocatello — Monida Pass
124 miles

Except in the area near Monida Pass, all the rocks exposed along this highway are volcanic, all part of the Snake River Plain. Monida Pass is in the western end of the Centennial Range, a complex assortment of folded and faulted sedimentary formations intruded by granite.

Fault Block Mountains

Mountains that fringe the southeastern side of the Snake River Plain make a ragged backdrop low along the eastern horizon between Pocatello and Roberts. They are fault blocks, part of the Basin and Range.

Rocks in those mountains include a variety of Paleozoic and Mesozoic sedimentary formations that were tightly folded and much broken along faults as the Rocky Mountains formed, about 70 to 80 million years ago. Erosion then reduced those mountains to what must have been an inconspicuous tract of low hills. In the last few million years, rhyolite ash blanketed large parts of them as the Yellowstone hotspot burned its way through the eastern part of the Snake River Plain. Active faults are now raising those mountain blocks and dropping the valleys between them.

Age dates show that about six million years ago a large rhyolite ash flow spread across part of the Blackfoot Mountains, which sprawl low along the horizon east of Blackfoot. That happened just as the eastern edge of the Basin and Range reached this area. Movement on those faults has since raised the Blackfoot Mountains at least 2000 feet, the

Section east of the highway, north of Pocatello.

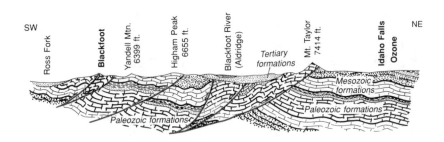

255

vertical distance between scattered patches of rhyolite on top of the range and on the nearby lowland across the faults along the northern edge of the range. Younger volcanic rocks around the Blackfoot Mountains are basalt lava flows erupted long after the rhyolite.

The Northern Rocky Mountains

North of Dubois, the Big Lost River, Lemhi, Beaverhead, and Centennial ranges, reciting from west to east, rise on the northern horizon. They consist mostly of Paleozoic and Mesozoic formations that were tightly folded and faulted as the Rocky Mountains formed. North of Spencer, the highway crosses granite and folded sedimentary rocks that belong to those mountains. All these ranges are certainly part of the Northern Rockies. All except the Centennial Range are also Basin and Range fault blocks, slices of the Northern Rockies that are pulling away from each other as movement on their bounding faults makes the valleys between them wider.

The Centennial Range, the one closest to the Yellowstone Plateau, is exceptional in trending west, almost at a right angle to the persistent north trend of the Basin and Range fault blocks. Perhaps the Centennial Range is high because it is on the edge of the Yellowstone Plateau, an area that bulges simply because heat from the hotspot in the mantle swells the mantle and continent crust. If so, then the area will drop as the hotspot moves east, and the crust in this region cools. In any case, Basin and Range faults will eventually dissect the bulge into more or less north-trending ranges and valleys as the hotspot migrates east.

The Snake River Plain

Between Roberts and Spencer, the highway steers an almost perfectly straight course directly across the nearly level surface of the eastern Snake River Plain. None of the volcanic rocks along this part of the road are more than a few million years old; some are just a few thousand years old.

Except for some rhyolite in the Spencer area, all the easily visible volcanic rocks are basalt lava flows, which certainly cover much greater volumes of older rhyolite. The rhyolite erupted as the Yellowstone hotspot passed, and the basalt came later as the Basin and Range faults began to break the Snake River Plain.

The volcanic vents that so liberally freckle the geologic map don't do much for the landscape. They are very small basalt shield volcanoes, low and inconspicuous rises a mile or so across and a few hundred feet high. You will see them if you look for them.

Pressure ridge on the Hell's Half Acre basalt flow.

The Hell's Half Acre Lava Field

The rest area 30 miles north of Pocatello provides a good place to walk around on the eastern end of the large Hell's Half Acre lava flow. It erupted from a small shield volcano that lies about 15 miles northwest of the rest area, well out of sight from the road. The flow covers most of the area between the road and the volcano — quite a few half acres. The abrupt transition from scabby basalt to wheat fields about two miles north of the rest area marks the edge of the flow.

Several excellent pressure ridges around the rest area make it easy to imagine how pieces of the hard crust on the cooling flow rafted along on still molten lava beneath, then jammed together to make the pressure ridge. Peek into the cracks in their crests to see places where gooey masses of molten lava squeezed between slabs of hard crust to make molded globs of basalt that look almost like black taffy.

Who would guess to look at this perfectly fresh flow almost without soil or plant cover that it could be all of 20,000 years old? Even though it is so much older than it looks, the Hell's Half Acre flow is much too young to be numbered among the volcanic offspring of the Yellowstone hotspot. This part of the Snake River Plain passed well beyond the probable reach of that torch several million years ago. The Hell's Half Acre flow is the child of Basin and Range volcanism, related to the other Basin and Range volcanic centers scattered here and there across southern Idaho.

The Hell's Half Acre lava flow, looking north along I 15.
—Bill Hackett photo

Precious Opal

You can dig your own gems, for a fee, at the Spencer Opal Mine, seven miles east of Spencer. The opals are outstanding, full of surprising flashes of red, yellow, blue, and green in absolutely pure spectral hues. Most of the stones are pale; a few are dark. Mounting a thin slice of naturally pale opal on a dark base makes the bright flashes appear to come from a black background.

Opal, one of the many varieties of silica, is a common mineral in many kinds of volcanic rocks. It typically forms nondescript masses of dull yellowish and greenish material that fill formerly open spaces in the rock. Precious opal with its bright flashes of colored light is extremely rare, and anything but nondescript.

All opal forms as a gelatinous precipitate of silica gel eventually hardens into the solid mineral without forming crystals. Precious opal forms where silica gel precipitates very slowly from absolutely quiet water that fills holes deep in the volcanic rock. Under those conditions, the silica gel coagulates from solution into microscopic spheres that settle very gently to the floor of the cavity, where they align themselves in regular layers and rows, like so many marbles shaken into a box.

When light strikes precious opal, the perfectly ordered layers of little spheres act as a natural diffraction grating to break the light up into the spectrum of different wavelengths. The precious opal separates the colors of light as though it were full of miniature prisms. Notice that the colored lights that flash from opals are the exact colors of the rainbow — no pastel shades, no mixed or muddy hues.

258

Basalt lava flow near the highway southeast of Boise.

Interstate 84:
Ontario — Mountain Home
94 miles

The Columbia Plateau and the Snake River Plain meet in the general vicinity of Boise. The western part of the route, between Boise and Ontario, crosses the Columbia Plateau on basalt erupted about 15 million years ago and on sediments deposited in lakes impounded behind those lava flows. The eastern part of the route, between Boise and Mountain Home, crosses much younger basalt flows and sediments that belong to the Snake River Plain.

Even though they differ fundamentally, the Snake River Plain and Columbia Plateau resemble each other in having basalt surfaces. They look more alike than they really are. Nevertheless, you can generally recognize Columbia Plateau basalts because they tend to

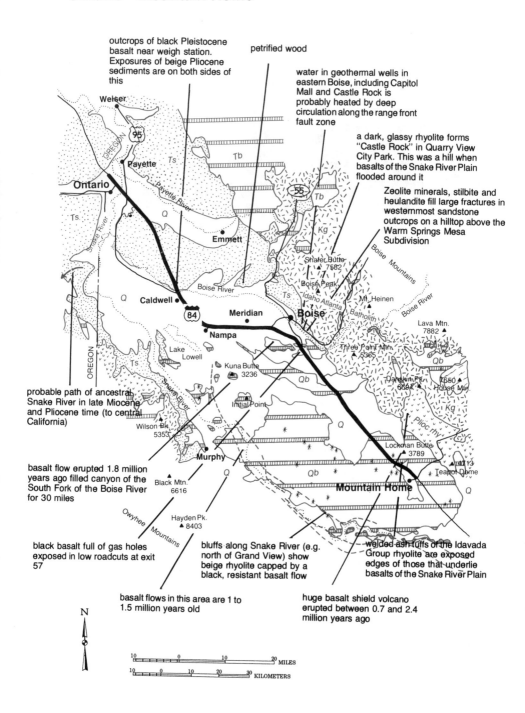

outcrops of black Pleistocene basalt near weigh station. Exposures of beige Pliocene sediments are on both sides of this

petrified wood

water in geothermal wells in eastern Boise, including Capitol Mall and Castle Rock is probably heated by deep circulation along the range front fault zone

a dark, glassy rhyolite forms "Castle Rock" in Quarry View City Park. This was a hill when basalts of the Snake River Plain flooded around it

Zeolite minerals, stilbite and heulandite fill large fractures in westernmost sandstone outcrops on a hilltop above the Warm Springs Mesa Subdivision

Weiser

95

Payette

Ts

Tb

Ontario

Payette River

Q

55

Tb

Kg

Emmett

Shafer Butte
▲ 7582

Boise Mountains

Boise Peak

Boise River

Caldwell

Q

Boise River

Ts

Idaho Atlanta

Mt. Heinen

84

Meridian

Boise

Batholith

Boise River

Nampa

Three Point Mtn.
▲ 5365

Lava Mtn.
7882 ▲

Lake Lowell

Ts

Kuna Butte
▲ 3236

Qb

Qb

probable path of ancestral Snake River in late Miocene and Pliocene time (to central California)

Snake River

Initial Point

Danskin Pk.
6684 ▲

7650 ▲
House Mtn

Q

Kg

Wilson Pk.
5353

Plioc rhy

Murphy

Lockman Butte
▲ 3789

basalt flow erupted 1.8 million years ago filled canyon of the South Fork of the Boise River for 30 miles

Black Mtn.
6616

Q

Qb

Teapot Dome

Mountain Home

Q

Owyhee Mountains

Hayden Pk.
▲ 8403

black basalt full of gas holes exposed in low roadcuts at exit 57

bluffs along Snake River (e.g. north of Grand View) show beige rhyolite capped by a black, resistant basalt flow

welded ash tuffs of the Idavada Group rhyolite are exposed edges of those that underlie basalts of the Snake River Plain

basalt flows in this area are 1 to 1.5 million years old

huge basalt shield volcano erupted between 0.7 and 2.4 million years ago

N

10   0   10   20 MILES
10   0   10   20   30 KILOMETERS

have a rusty brownish color, in subtle contrast with the almost flat black of the Snake River Plain basalts.

The Owyhee Mountains rise in the distance southwest of the road. They consist basically of granite the same age as that in the Idaho batholith, late Cretaceous — probably the same granite. The thick blanket of late Miocene rhyolite ash that covers large areas of the granite erupted in the aftermath of the meteorite impact of 17 million years ago. Passage of the Yellowstone hotspot through this area in latest Miocene time isolated the granite in the Owyhee mountains from the main mass of the Idaho batholith.

Mountains northeast of the road between Boise and Mountain Home are mostly granite of the Atlanta batholith, the southern part of the Idaho batholith. Miocene rhyolite, part of the Idavada group of volcanic rock formations, covers the granite in the mountains directly north of Mountain Home. The Idavada group is about 15 million years old, a little too old to be part of the Snake River Plain, the right age to associate it with the Columbia Plateau.

Columbia Plateau

Between Ontario and the Nampa area, the road crosses the southern part of the Weiser embayment, an eastern lobe of the Columbia Plateau. You would hardly suspect it from the small amounts of basalt exposed near the road. The few outcrops that do appear expose sediments, most of which were deposited in Miocene lakes impounded behind the lava flows. Sediments sandwiched between the lava flows record the existence of at least several such lakes, some of which may have been fairly large. But none could have been deeper than the thickness of the lava flow that dammed it, probably less than 200 feet. Imagine the trees swaying along their shores in the mild and wet climate Idaho enjoyed during that part of late Miocene time.

Although the name is rarely used in this part of Idaho, the sediments that accumulated in those lakes are equivalent to the Latah formation, which records the existence of similar lakes farther north. The lake sediments contain abundant plant fossils, mostly dark leaf impressions on layers of clay or rhyolite ash, but very few bones or pieces of petrified wood. That is a real deprivation. It would be fascinating to see more remains of the fish that swam in the lakes, of the animals that lived along their shores, and to have a few solid chunks of the forest. Sad as it is, the scarcity of such fossils is easy to understand.

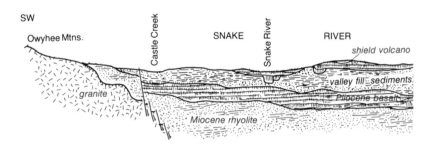

SW

Owyhee Mtns.

Castle Creek

SNAKE

Snake River

RIVER

shield volcano

granite

valley fill sediments

Pliocene basalt

Miocene rhyolite

Section across the line of the highway

Decaying organic matter generally makes freshwater lakes in heavily forested regions slightly acidic. We think it is entirely possible that the intense volcanic activity in the Columbia Plateau caused strongly acid rains that might have made those lakes more than slightly acidic, especially in the months or years after the big flood basalt eruptions.

Unfortunately for people who enjoy fossils, the mineral matter in bones is calcium phosphate, which readily dissolves in weak acids. That explains the shortage of animal bones; they simply dissolved in the weakly acidic water as though they were so many sugar cubes dropped into a cup of hot coffee. Acidity also prevented wood from petrifying. Silica, the mineral matter that petrifies wood, does not dissolve in acidic water, so no dissolved silica was available to saturate and petrify the wood before it could decompose.

While they dissolved bones, the acidic lake waters also bleached the sediments by dissolving iron oxide. That converted the red mud that washed into the lake into white kaolinite clay, which now provides raw material for porcelain and fire brick. If the water had not been acidic, the lake sediments would be red or brown.

Fish fossils do survive in the younger late Miocene lake sediments deposited in the Snake River Plain after the wet climatic period ended. They so closely resemble fish fossils from California that some geologists believe a river flowed from Idaho to California during late Miocene time. How else could the fish look so much alike? So far, no one has recognized remnants of an old valley in the modern Basin and Range landscape, but it would be extremely difficult to recognize a thoroughly defunct valley in a region so thoroughly chopped up. Regardless of any possible California connection, we think one of the large lakes impounded behind the lava flows overflowed north to establish the spillway that eventually became Hells Canyon.

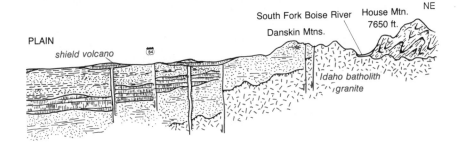

about six miles northwest of Mountain Home.

Western Snake River Plain

Imagine the Snake River Plain as an arm bent at an elbow near Twin Falls; the forearm extending northeast to a long finger laid on Yellowstone Park, the upper arm reaching northwest. This highway crosses nearly the full length of the upper arm, the western Snake River Plain. It has nothing to do with the migrating hotspot track that created the main part of the Snake River Plain.

The western Snake River Plain trends northwest, and its margins appear to be Basin and Range faults. Very little hard bedrock appears at the surface. Deep drill holes reveal that the rocks beneath the surface consist mostly of valley fill and lake bed sediments, basalt lava flows, and rhyolite — all complexly interbedded. One well drilled about 50 miles south of Boise went through approximately two miles of such stuff before it finally penetrated granite.

These flattened gas bubbles exposed in a road cut in basalt near Boise were probably round when they formed, then stretched flat as the flow continued to move.

263

The simplest view of the western Snake River Plain is to interpret it as a Basin and Range fault block that acquired a deep fill of volcanic and sedimentary rocks as it dropped. Age dates on volcanic rocks and fossils in the sedimentary rocks both show that the western Snake River Plain formed during the last ten or so million years. Some of the basalt flows are less than a half million years old. Like the main part of the Snake River Plain, this western part seems to consist mostly of rhyolite, which generally lies beneath a cover of basalt flows. The rhyolite is ash that drifted in from volcanoes in the main part of the Snake River Plain; the flows accompanied Basin and Range faulting.

Natural Hot Water

Ever since the last century, some public buildings and a few private homes in Boise have used natural hot water from deep wells for space heating. Those of us who worry about utility bills during the winter months must envy people who can tap an apparently endless supply of hot water simply by drilling a well.

As in so many hot springs, the steaming water beneath Boise appears to have gained heat as it circulated to depths of several thousand feet in the fracture zones along faults. Then, porous sandstone and fractured volcanic rocks a thousand or more feet below the surface trap the rising hot water. The natural insulating properties of rock conserve the heat.

Crater Rings

A nicely matched pair of nearly circular craters glares like two staring eyes from maps of the area about eight miles northwest of Mountain Home. The western crater is about 2500 feet across and 200 feet deep; its neighbor, a few hundred feet to the east, is about 2800 feet across and 250 feet deep. Their origin is not completely clear. The simplest explanation has them forming in great steam explosions that happened after molten basalt magma rose into water saturated rock beneath the surface of the Snake River Plain.

Basalt lava flow on the pale sediments of the Glenns Ferry formation and rhyolite. —U.S.Geological Survey photo by H.E. Malde

Interstate 84:
Mountain Home —
Junction Interstate 86
105 miles

The entire route is on the Snake River Plain. Bedrock consists entirely of basalt lava flows at the surface, rhyolite ash at shallow depth beneath the basalt. Here and there, cliff exposures that show a thin cap of black basalt over a thick section of pale rhyolite illustrate that relationship.

The small shield volcanoes that erupted the basalt lava flows make occasional low mounds a mile or so across that rise inconspicuously above the Snake River Plain north of the highway. Watch carefully for them, especially near Hammett or Glenns Ferry, and between Twin Falls and Burley. Their gentle bulges in the landscape are easy to overlook.

Section across the line of the highway near Bliss.

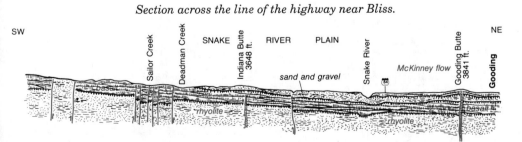

265

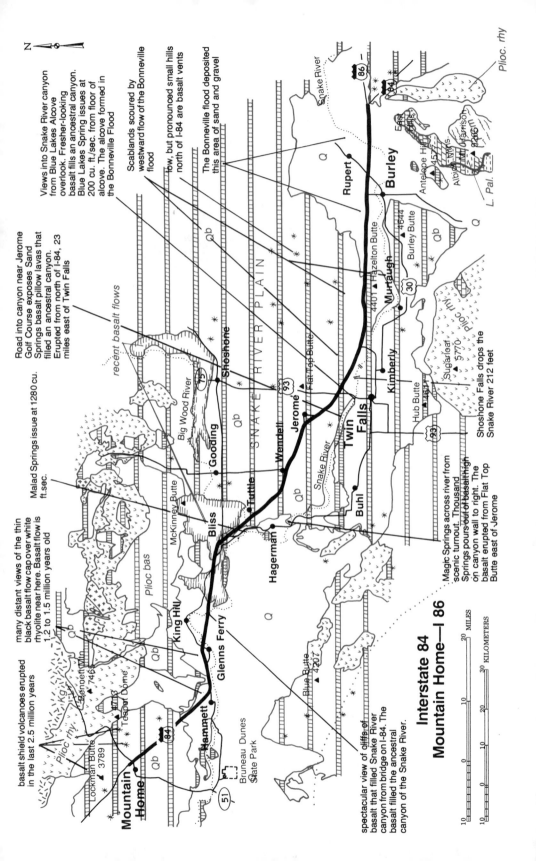

Interstate 84
Mountain Home—I 86

N

basalt shield volcanoes erupted in the last 2.5 million years

many distant views of the thin black basalt flow cap over white rhyolite near here. Basalt flow is 1.2 to 1.5 million years old

Malad Springs issue at 1280 cu. ft./sec.

Road into canyon near Jerome Golf Course exposes Sand Springs basalt pillow lavas that filled an ancestral canyon. Erupted from north of I-84, 23 miles east of Twin Falls

Views into Snake River canyon from Blue Lakes Alcove overlook. Fresher-looking basalt fills an ancestral canyon. Blue Lakes Spring issues at 200 cu. ft./sec. from floor of alcove. The alcove formed in the Bonneville Flood

Scablands scoured by westward flow of the Bonneville flood

flow, but pronounced small hills north of I-84 are basalt vents

The Bonneville flood deposited this area of sand and gravel

recent basalt flows

Plioc. bas

Plioc. rhy.

Qb

Q

S N A K E   R I V E R   P L A I N

Big Wood River

McKinney Butte

Bennett Min
▲ 7466

Teapot Dome

Lockman Butte
▲ 3789

Plioc. rhy.

Kg

Qb

Mountain Home

Hammett

84

51

Bruneau Dunes State Park

Glenns Ferry

King Hill

Blue Butte
▲ 4207

Q

Hagerman

Bliss

Tuttle

Gooding

Wendell

Shoshone

75

93

Jerome

Buhl

Twin Falls

Kimberly

Snake River

Hub Butte
▲ 4611

93

Flat Top Butte

Murtaugh

30

Hazelton Butte
4401 ▲

Burley Butte
▲ 4644

Qb

Rupert

Burley

86

Snake River

Snake River

Q

Antelope Hills
▲ 4457

Albion Mts.
2565 m

East Hills

L. Pal.

Plioc. rhy.

Plioc. rhy.

Sugarloaf
▲ 5770

spectacular view of cliffs of basalt that filled Snake River canyon from bridge on I-84. The basalt filled the ancestral canyon of the Snake River

Magic Springs across river from scenic turnout. Thousand Springs pours out of basalt high on canyon wall to right. The basalt erupted from Flat Top Butte east of Jerome

Shoshone Falls drops the Snake River 212 feet

0   10   20 MILES

0   10   20   30 KILOMETERS

Except for a short stretch east of Glenn's Ferry, most of the route lies close to the north edge of the Snake River Canyon. That deep cut provides occasional glimpses of rocks that would otherwise lie completely hidden and unsuspected beneath the basalt. The canyon has been the scene of some strange events.

McKinney Lake

Along much of the route between Glenns Ferry and the area just east of Bliss, the highway crosses the McKinney basalt flow. It poured south and west from McKinney Butte, a volcano about nine miles northeast of Bliss, and filled the Snake River Canyon to impound ancient Lake McKinney. The water was 600 feet deep at the lava dam, and flooded the canyon upstream beyond Thousand Springs, well into the Melon Valley.

The McKinney Basalt Flow.

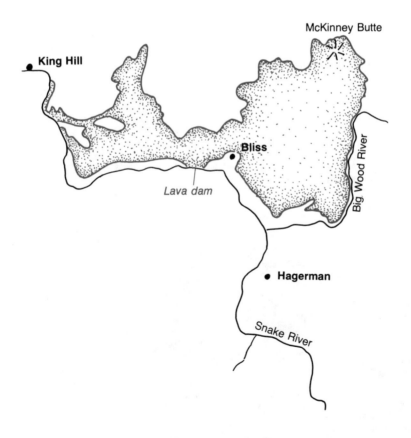

Lake McKinney lasted a long time because its only outlet was slow seepage through and beneath the lava flow, a kind of outflow that does not erode a deep spillway to drain the lake. Deep deposits of clay that slowly filled the lake finally sealed the lava dam tight, forcing the water to overflow and develop a spillway across the Glenns Ferry formation south of the old canyon. Rapid erosion of the soft sediments then carved the spillway into a new canyon that drained the lake, while leaving the lava dam intact. Scattered outcrops expose the basalt that dammed the old canyon. It is mainly pillow basalt because most of the flow poured into the water backing up behind the dam as it formed.

The Hagerman Zebra Herd

The bluff across the canyon from Hagerman nicely exposes the Glenns Ferry formation. It is a thick pile of sedimentary deposits laid down during the long dry spell of Pliocene time, the usual assortment of sediments deposited in a desert valley, including some lake beds.

Several paleontologists quarried quite a deep cut in the bluff as they collected skeletons of dozens of horses, along with an amazing variety of other animals, from a layer of sediment about 450 feet above the river. Age dates on basalt lava flows below the bone bed, and on some rhyolite ash above it, show that the animals lived about three million years ago.

Some of the paleontologists who dug up the bones thought the situation may have resembled that at a modern African water hole at the end of a long drought, where desperately thirsty animals bog down in the mud, then fall prey to fierce predators that risk the same fate. Certainly the evidence suggests that Idaho was a terribly dry place three million years ago, where animals might well have risked everything to drink at a water hole.

Sediments of the Glenns Ferry formation exposed near Glenns Ferry.
— U.S. Geological Survey photo by H.E. Malde

268

To judge from the skeletons in the Glenns Ferry formation,, the horses that roamed the Snake River Plain three million years ago were almost identical to the modern zebras of Africa. No one can ever know whether Idaho's zebras had stripes; the total lack of evidence leaves you perfectly free to imagine them wearing spots or any other ornamental markings.

The Weeping Canyon Wall
— Thousand Springs

The north wall of the Snake River canyon tends to weep nearly everywhere, especially in the Thousand Springs, which discharge about 600 cubic feet of water per second. Other especially notable seepages include Malad, Blue, Sand, and Niagara springs.

Years ago, spring water poured down the north canyon wall in a long series of magnificent cascades. Diversion of water to power generation, irrigation, and fish hatcheries has since destroyed much of the original natural beauty. Nevertheless, enough remains here and there to help you imagine what it was like when Idaho was young. Niagara Springs, about 20 miles west of Twin Falls, is probably the best preserved of the lot.

Section through the Snake River Plain showing how the canyon intercepts water percolating south through the lava flows.

N   S

springs

basalt flows

Snake River

All those springs exist because basalt lava flows thirstily absorb surface water, and the Snake River Plain has a very gentle southward tilt. Water draining from the Rocky Mountains of central Idaho soaks into the basalt lava flows along the north edge of the Snake River Plain, then slowly percolates southward through the porous zones of rubble sandwiched between the flows. The water leaks out of the basalt flows in countless springs along the north edge of the Snake River canyon. So streams such as the Big Lost River that sink so hopelessly into the basalt along the north edge of the Snake River Plain really are tributaries of the Snake River, even though they don't manage to flow very far across the absorbent volcanic surface.

Malad Gorge

The Big Wood River flows through Malad Gorge to enter the Snake River just north of Hagerman. Malad Gorge is obviously much too large to have been eroded by the small river that now flows through it, and much carved into a complex of bedrock terraces, alcoves, and balconies. Black boulders of the melon gravel litter many of the flatter surfaces. In general, Malad Gorge looks like the familiar handiwork of the Bonneville flood rushing through a narrow canyon.

Like many places on the north side of the Snake River, Malad Gorge is full of large springs. Much of the flow is now diverted to power generation and irrigation, but the springs that remain in something resembling their natural condition are still worth a visit.

The Gooding City of Rocks

The Gooding City of Rocks is actually about 12 miles north of Gooding. The easiest part of the area to visit is the Little City of Rocks. Take Idaho 46 north from Gooding; travel a bit more than 12 miles,

Rhyolite cliffs north of Gooding. —U.S. Geological Survey photo by C.P. Ross

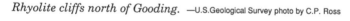

then watch for a sign pointing the way another mile west on a dirt road. That leads you directly to one of eight canyons eroded into rhyolite ash in the Bennett Hills. The other canyons are within hiking distance.

Eruptions of several million years ago laid down deep deposits of rhyolite ash, some more tightly welded than others. Weathering and erosion have since preferentially attacked the less resistant rocks, leaving the harder material standing in bold relief. The result is a fairy land of grotesque rock sculptures, a great place for an unusual hike amid splendid exposures of rhyolite ash.

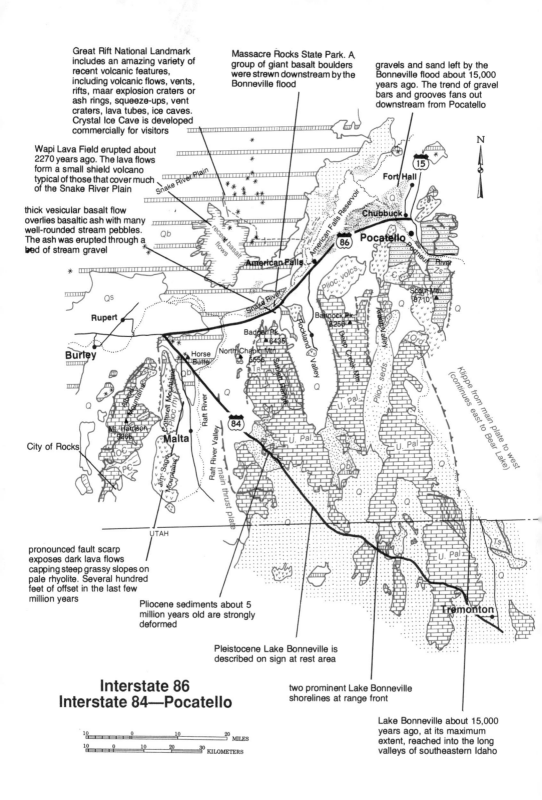

Great Rift National Landmark includes an amazing variety of recent volcanic features, including volcanic flows, vents, rifts, maar explosion craters or ash rings, squeeze-ups, vent craters, lava tubes, ice caves. Crystal Ice Cave is developed commercially for visitors

Massacre Rocks State Park. A group of giant basalt boulders were strewn downstream by the Bonneville flood

gravels and sand left by the Bonneville flood about 15,000 years ago. The trend of gravel bars and grooves fans out downstream from Pocatello

Wapi Lava Field erupted about 2270 years ago. The lava flows form a small shield volcano typical of those that cover much of the Snake River Plain

thick vesicular basalt flow overlies basaltic ash with many well-rounded stream pebbles. The ash was erupted through a bed of stream gravel

pronounced fault scarp exposes dark lava flows capping steep grassy slopes on pale rhyolite. Several hundred feet of offset in the last few million years

Pliocene sediments about 5 million years old are strongly deformed

Pleistocene Lake Bonneville is described on sign at rest area

two prominent Lake Bonneville shorelines at range front

Lake Bonneville about 15,000 years ago, at its maximum extent, reached into the long valleys of southeastern Idaho

N

Fort Hall

Chubbuck

Pocatello

American Falls

Snake River Plain

American Falls Reservoir

Scout Mtn 8710

Rupert

Burley

Bannock Pk 8250

Badger Pk 6435

North Chapin Mtn 5555

Horse Butte

Rockland Valley

Deep Creek Mtns

Arbon Valley

Plioc. seds

Klippe from main plate to west (continues east to Bear Lake)

Raft River

Raft River Valley

Malta

City of Rocks

Mt. Harrison 9265

Cotterell Mountains

Jim Sage Mountains

main thrust plate

U. Pal.

U. Pal.

U. Pal.

U. Pal.

Tremonton

UTAH

Interstate 86
Interstate 84—Pocatello

10   0   10   20 MILES

10   0   10   20   30 KILOMETERS

Interstate 86:
Interstate 84 — Pocatello
62 miles

The highway follows the south side of the Snake River Canyon all the way, with the volcanic Snake River Plain stretching away to the north, and Basin and Range mountains rising along the southern horizon. All the bedrock along the road is basalt, best exposed in the steep canyon walls. The canyon bears the mark of the Bonneville flood in its eroded walls and big deposits of flood sediments.

Fault Block Mountains

From west to east, the mountains and valleys south of the highway include the Sublett Range southwest of American Falls, the Rockland Valley almost directly south of town, and the Deer Creek Range slightly east of south. The Arbon Valley and Bannock Range are much farther east.

Section through the ranges and valleys south of American Falls.

As everywhere in Idaho, the steeply dipping Basin and Range faults chop across the much older structures of the northern Rocky Mountains, in which the faults lie nearly flat. Our section also shows a series of fault blocks beneath the valley floors. Those are buried and therefore to some extent conjectural, but similar structures exist here and there on the flanks of the mountains, where you can see them. Basin and Range valleys differ from the mountains only in being low, and flooded with sediment. If the climate of this region were wetter, streams would empty the sediment out of those valleys.

273

Flushing the Canyon

The Bonneville flood left its mementos in nearly every part of the Snake River Canyon, including near the highway: dramatic erosion where the water rushed through narrow gorges, spectacular deposition in the more spaciously broad reaches of the canyon where the water spread out to flow more slowly. Their regular patterns of vertical fractures make basalt lava flows terribly vulnerable to flood erosion.

Flood water roiling through the narrow reaches of the canyon plucked the basalt columns apart, and swept the pieces along the canyon floor. You can see the effects of that erosion in the numerous blind alcoves along the canyon walls, each probably marking the former site of a rapidly spinning whirlpool. The plucked blocks of basalt rolled through the narrow parts of the canyon, then landed in the big deposits of sediment, giant gravel bars laid down just where the water slowed as the canyon widened. Few of the blocks of basalt travelled more than a mile or two, just enough rolling to round off their sharp edges and corners, shape them into the black watermelon boulders that litter those giant river bars.

As the flood began to subside, the round boulders stopped moving first, while the slowing current still carried finer sediments. The lagging basalt boulders armored the surfaces of the bars, protecting the fine sediments beneath. If you dig a hole into one of those bars, you will generally find much finer sediment beneath the bouldery surface.

An intricate pattern of layers in flood deposited sand.
— U.S. Geological Survey photo by H.E. Malde

American Falls Lake and the Bonneville Flood

American Falls Reservoir was first filled in 1927 to provide irrigation water for a large expanse of formerly dry farmland. It is the partial restoration, so to speak, of a much larger natural lake that once existed in the same area. About 72,000 years ago, give or take about 14,000 years, a large lava flow that erupted from Cedar Butte volcano nine miles southwest of American Falls dammed the Snake River. That flow impounded ancient American Falls Lake, which reached upstream almost to Blackfoot.

Some of the earlier sediments laid down in the lake contain abundant fossils, which include pollen, a variety of water weeds, and snails. Those older fossils tell us that the climate of some 60,000 or more years ago was very much like what we know today, perhaps slightly cooler and wetter. Evidently that was before the last ice age started. Younger fossils higher in the sedimentary section show the climate becoming much cooler and wetter, probably as the most recent ice age began.

As the glaciers of that ice age reached their maximum sometime around 15,000 years ago, Lake Bonneville overflowed through Red Rock Pass to release the Bonneville flood down the valleys of Marsh Creek and the Portneuf River, then into the Snake River. The flood suddenly raised the lake level of ancient American Falls Lake, causing it to overflow and drain through a rapidly eroding spillway, now part of the Snake River canyon.

Along most of the way between American Falls and Chubbuck, the highway follows the south edge of an enormous debris fan dumped where the Bonneville flood poured out of the valley of the Portneuf River and into ancient American Falls Lake. Watch for gently sloping broad surfaces north, and in some places on both sides, of the road. They are underlain by flood deposits of sand and gravel, which becomes coarser as it approaches the mouth of the Portneuf River. Some of the basalt boulders near Pocatello are as much as eight feet across.

The Basalt Boulders at
Massacre Rocks State Park

Most of the bedrock exposed in Massacre Rocks State Park is a basalt agglomerate, an unsightly mess composed mainly of chunks of black basalt about the size of walnuts set in a fine-grained matrix that consists largely of dark gray basalt ash. That stuff is probably a volcanic mud flow. Imagine it forming as heavy rain soaked freshly fallen ash, converting it into a mudflow with the consistency of wet

Blocks of basalt lie where the subsiding waters of the Bonneville flood left them. An Indian fight that went badly for a group of travelers on the Oregon Trail gave the Massacre Rocks their name.

cement, that picked up chunks of basalt as it gurgled down the slope. Sometime later, a massive basalt lava flow covered the rubbly agglomerate.

When the Bonneville flood came this way in all its roaring fury, it ripped enormous chunks of basalt from the lava flow, and swept them several hundred yards downstream, where it dumped them on exposures of the agglomerate. Now those blocks lie scattered across the park, their slightly rounded edges and corners clearly showing that they rolled under the floodwaters. Had they rolled much farther, those edges and corners would be more than slightly rounded.

VOLCANOES ALONG THE GREAT RIFT

The choicest geologic spectacles in this part of Idaho are a series of very young lava flows and other volcanic features in the Wapi and Kings Bowl lava fields north of the Snake River. The basalt looks so fresh and new and hot that it is difficult to believe that the youngest flows are a bit more than 2000 years old, actually cold. All the eruptions align along the Great Rift, a swarm of open fissures that cuts completely across the Snake River Plain.

To get to the Wapi and Kings Bowl lava fields and the Great Rift, cross the Snake River bridge at American Falls and follow the road to Aberdeen, then watch for the signs to Great Rift National Landmark and Crystal Ice Cave. That route will take you past the Wapi Lava field and into the Kings Bowl Lava field. The roads into Wapi Park and beyond Crystal Ice Cave are little more than tracks much too primitive for most cars, and most drivers. In this discussion, we will limit ourselves to several of the more accessible volcanic features. Many others await people interested in venturing farther.

Recent basalt flows erupted from the Great Rift.
Air view from the north. —Bill Hackett photo

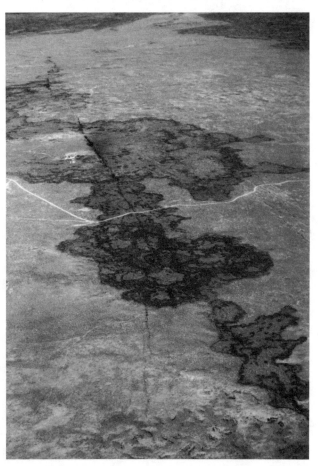

The view along one of the fissures of the Great Rift. The darker view straight down is truly appalling.

The Great Rift

The Great Rift is a world class geologic spectacular, a series of open fissures and swarms of open fissures that trend slightly west of north and break all the way across the eastern Snake River Plain. The fissures are the source and center line of a gallery of volcanic features most of which probably erupted within a short time.

The northern end of the rift system is in Craters of the Moon National Monument, where the fissures are filled with basalt. The rifts continue southeast through the Open Crack fissures, which extend to depths of at least 800 feet — climbers who descended that far found ice at the bottom. You can stand on the edge and stare into the black depths. Farther southeast are the Kings Bowl fissures, which produced the Kings Bowl lava field. The southernmost fissures erupted the Wapi lava field, about 20 miles west of American Falls.

That long swarm of open fractures with their general northwest orientation shows beyond any shadow of a doubt that movement along Basin and Range faults is stretching the earth's crust. Their extreme youth precludes any idea of attributing the fractures or the lava flows to the Yellowstone hotspot, which must have left this part of the Snake River Plain in its wake several million years ago. These are younger volcanic features superimposed on an older volcanic province, not a late stage of activity associated with the Yellowstone hotspot.

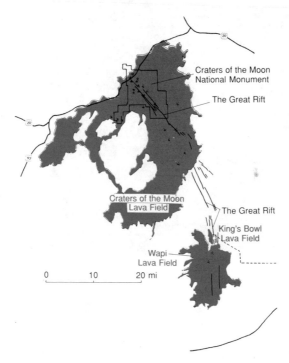

*The sets of fractures that together make the southern part of the
Great Rift system, and the lava fields that erupted from them.*

The Kings Bowl Lava Field

In the volcanic scale of things, Kings Bowl crater is a minor feature,
a wide spot on the Great Rift almost 100 feet deep and about the size
of a football field. That shallow crater is the main vent from which
several thin flows erupted to make the surrounding Kings Bowl lava
field, which covers an area of about one square mile. Radiocarbon
dates on lava-burned sagebrush collected from beneath the flows
show that they erupted 2130 years ago, give or take a century.

Crystal Ice Cave, within the Kings Bowl lava field, is a part of the
King's Bowl Rift that contains large masses of ice, much of it in bizarre
and beautiful formations. Ice caves are strange in being much colder
than normal caves, which typically stay near the local mean annual
temperature all year round.

Ice typically fills caves that open vertically to the surface, with no
cross-ventilation. Only air that is denser, in other words colder, than
that already inside can sink into such a cave. So air enters the cave

Looking into Great Rift in the Kings Bowl area.

only during the winter, never on warm summer afternoons. The rock walls of the cave insulate it well enough to maintain those cold temperatures all year, keeping the cave cold, and full of ice.

The Wapi Lava Field

The southern end of the Great Rift system erupted the large Wapi lava field. The flows look so fresh that you almost expect to see them glow red at night as the ground rumbles with escaping steam. It is difficult to believe these lavas could be as much as 2000 years old.

The Wapi lava field is essentially a shield volcano composed of a pile of basalt flows that erupted from a central vent, one after another. The big stump of Pillar Butte, a prominent landmark visible throughout this part of the Snake River Plain, marks the vent about the way a flag marks the hole on a golf green. Pillar Butte is a large spatter cone, a chimney of lava that formed as escaping steam coughed up globs of molten basalt, which stuck together as they cooled.

Split Butte

Split Butte eventually rewards those who navigate the difficult road beyond Crystal Ice Cave. It is among the older and more complex volcanoes in the southern part of the Great Rift system. Activity began as roaring jets of red hot steam blasted clouds of molten shreds of basalt out of the vent. The basalt cinders settled in a broad ring around the vent to make a pile shaped approximately like a doughnut. Imagine the spectacular view at night, as the flying globs of molten basalt traced glowing red arcs in the sky, then dropped dimming into the growing wall of cinders.

When most of the steam had blown off, dry lava rose and quietly filled the basin enclosed within that doughnut ring of basalt cinders to form a nearly circular lake of lava. After a solid crust had formed on the lava lake, the still molten basalt within drained back into the fissure. That allowed the center of the crust to collapse, forming the broad pit in the middle of Split Butte. But the edges were already solid enough to remain standing as the shelf that surrounds the central pit.

After Split Butte formed, flows of the Wapi lava field approached it, but could not move up the gentle slope of the outer ring of cinders. So the steep margins of those younger flows form a wall that encircles Split Butte, making it look as though it has a sort of moat around it.

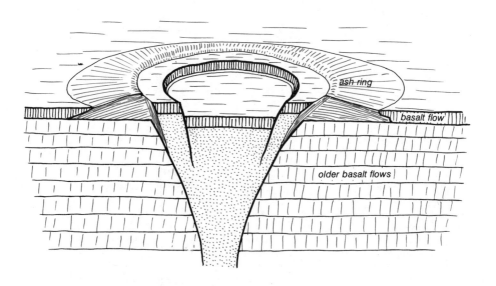

A schematic diagram of Split Butte.

US 20
Mountain Home—Arco

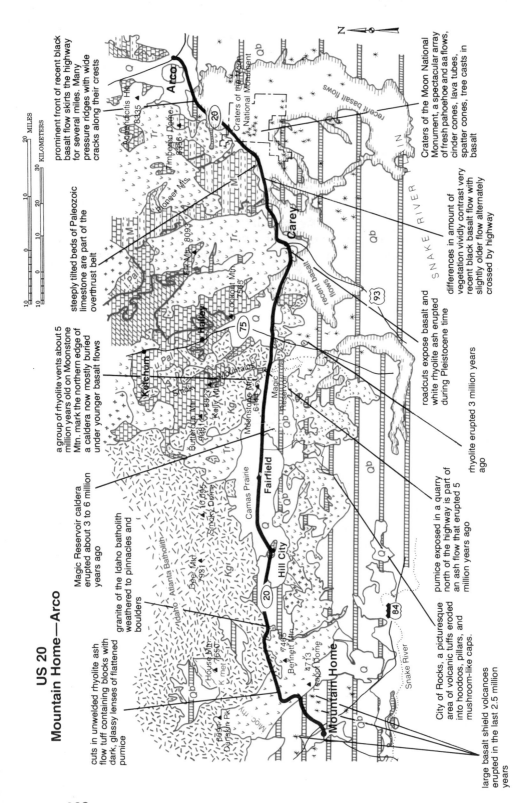

cuts in unwelded rhyolite ash flow tuff containing blocks with dark, glassy lenses of flattened pumice

Magic Reservoir caldera erupted about 3 to 6 million years ago

a group of rhyolite vents about 5 million years old on Moonstone Mtn. mark the northern edge of a caldera now mostly buried under younger basalt flows

steeply tilted beds of Paleozoic limestone are part of the overthrust belt

prominent front of recent black basalt flow skirts the highway for several miles. Many pressure ridges with wide cracks along their crests

granite of the Idaho batholith weathered to pinnacles and boulders

roadcuts expose basalt and white rhyolite ash erupted during Pleistocene time

pumice exposed in a quarry north of the highway is part of an ash flow that erupted 5 million years ago

City of Rocks, a picturesque area of volcanic tuffs eroded into hoodoos, pillars, and mushroom-like caps.

large basalt shield volcanoes erupted in the last 2.5 million years

rhyolite erupted 3 million years ago

differences in amount of vegetation vividly contrast very recent black basalt flow with slightly older flow alternately crossed by highway

Craters of the Moon National Monument, a spectacular array of fresh pahoehoe and aa flows, cinder cones, lava tubes, spatter cones, tree casts in basalt

282

Highway 20 follows close to the north edge of the Snake River Plain all the way between Mountain Home and Arco, cutting back and forth between young volcanic rocks and the much older formations of the northern Rocky Mountains.

In one stretch about midway between Mountain Home and Hill City, the road leaves the flat volcanic surface to cut through the mountains across about ten miles of granite. The road crosses more granite in the area north of the Magic Reservoir. All the rest of the bedrock exposed along and near this road is volcanic rocks that belong to the Snake River Plain: black basalt and pale rhyolite. Watch for small cinder cones, which generally perch on the broad, gentle domes of extremely inconspicuous shield volcanoes. They mark the volcanic vents.

Basalt and Rhyolite

The older volcanic rocks, mostly basalt lava flows, along the western part of the route lie beneath a fairly complete cover of soil, stream sediments, wind blown dust, and plants. You don't see much solid rock. The stream sediments washed in from the mountains to the north, then stalled on the Snake River Plain as the water that carried them sank into the basalt. Those deposits along its northern edge partly explain why the surface of the Snake River Plain slopes gently southward, why most of the basalt lava that erupts along the north edge of the Snake River Plain flows south.

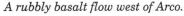

A rubbly basalt flow west of Arco.

The eastern part of the route crosses large tracts of very young basalt flows; some of those in the Craters of the Moon area are only about 2000 years old. The younger flows are almost perfectly bare, jagged seas of rough basalt still as morbidly black as the day it solidified. Only a few pioneering plants survive on the harsh surfaces of those young lava flows; their roots maintain a precarious hold by following obscure fractures that provide an occasional sip of water.

The Boise River Versus the Lava Flows

About 40 miles east of Boise, just north of the Snake River Plain and a few miles north of the highway, the Boise River and its South Fork flow through deep canyons eroded in granite of the southern part of the Idaho batholith, the Atlanta batholith. Volcanoes have greatly complicated the river's routine task of canyon cutting.

A series of basalt lava flows erupted from nearby volcanoes north of the Boise River filled those canyons time and again, each time setting the river back in its efforts to erode the canyon. The oldest flows are about two million years old, the youngest close to 200,000 years. Some of the flows are very large: The Steamboat Rock flow, for example, followed the river almost 40 miles, nearly to Boise.

Section through the Magic Reservoir caldera

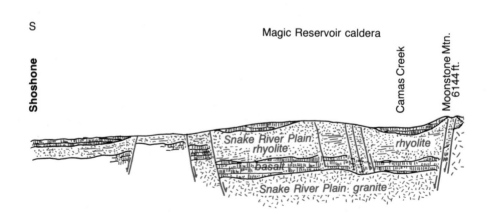

As each flow poured into the canyon, it immediately dammed the river to impound a lake. Then, basalt flowing into the rising lake broke up into cylindrical masses called pillows that range in size from buckets to barrels. The accumulating pillow basalts continued to raise the level of the lava dam until the eruption finally stopped. Some of the lava dams were hundreds of feet high. Like all dams, they impounded sediment as well as water.

Each lake finally filled with so much sediment that it could trap no more. Then, muddy water began to pour over the dam, eroding it and the lava flow downstream. An overflow of clear lake water from a lake that is still trapping sediment is not very effective in eroding granite or basalt. It is the sediment that running water carries, not the water itself, that scours hard rock.

Even with the aid of sediment, eroding solid rock is a terribly slow process. Geologists who studied the Boise River estimated that it takes the stream about a century to cut through an inch of basalt. After some 200,000 years, the river still has not cut to the base of the Smith Prairie flow — Smith Prairie, by the way, is a shield volcano.

These lava flows along the Boise River are much too young to associate with the passage of the Yellowstone hotspot. It passed well south of here sometime around six million years ago. They must belong to the Basin and Range.

and into the northern Rocky Mountains.

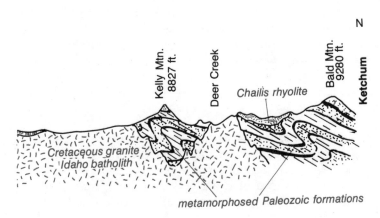

The Magic Reservoir Caldera

According to some geologists, the highway crosses an old caldera through the area about 10 miles west of Magic Reservoir. They are probably right. The complex and imperfectly known volcanic history of the Magic Reservoir caldera apparently started with the explosive eruption of large sheets of rhyolite ash about 10 million years ago, during late Miocene time. Those eruptions must have opened a caldera basin — it is hard to imagine how such enormous volumes of rhyolite ash could erupt without causing the earth's surface to subside into the space they empty.

After the big ash flow eruptions, large rhyolite lava flows and lava domes quietly filled the caldera. Those later eruptions may have continued until about three million years ago. Numerous basalt lava flows have since covered much of the countryside, effectively obscuring the outlines of the caldera. Nothing is quite as opaque as a basalt lava flow.

The Magic Reservoir caldera is certainly the handiwork of the Yellowstone hotspot. It was active at about the time this part of the Snake River Plain should have passed over the burner, and what little you can see of it looks like the ruins of a typical resurgent caldera.

Our geologic section leaps from direct observation to mild speculation in showing a large mass of Pliocene granite at shallow depth beneath the Magic Reservoir caldera. We feel confident that the granite really does exist despite the absence of exposed granite in the Snake River Plain. Granite and rhyolite have precisely the same composition, and differ in appearance only because one crystallized at depth while the other erupted. Granite intrusions are exposed by deep erosion of rhyolite in calderas elsewhere.

Millions of years from now, the slow processes of erosion will have stripped all the basalt and rhyolite off the surface of the Snake River Plain. We suggest that the revised geologic maps of that distant date will show a broad swath of granite cutting through the northern Rocky Mountains precisely where present maps show the broad volcanic sweep of the Snake River Plain.

The Elusive Antler Mountains

The Antler orogeny is one of the least understood mountain building events. It apparently happened during Devonian and Mississippian time, sometime around 350 million years ago, more or less. Evidently, the western margin of the North American continent collided then with the ocean floor, with the usual consequences. Most of the evidence of its happening exists in widely scattered localities in western Nevada and eastern California. Rocks in the mountains north of Carey appear to contain the best evidence of the Antler orogeny so far found in the Northern Rockies.

Those rocks include a sequence of sandstones, muddy sandstones, pebble conglomerates, and related rocks more than 4000 feet thick, the kind of sedimentary pile that normally accumulates in deep water near a continental margin. Widely scattered fossils suggest that the whole pile accumulated during Mississippian time. Now, those off-shore rocks lie on top of limestone of the same age that appears to have been deposited in very shallow water near shore. It seems that the thick section of offshore rocks was shoved onto the edge of the continent. It is fairly easy to imagine that sort of thing happening if the continent were to collide with the ocean floor. Remember that the western border of North America was then in western Idaho, not far from this area.

CRATERS OF THE MOON
NATIONAL MONUMENT

The Craters of the Moon lava field was the scene of the most recent volcanic eruptions in the Snake River Plain in which about 4.5 cubic miles of lava covered an area of approximately 643 square miles. That volume of lava is negligibly small by the extravagant standards of the Columbia Plateau, respectably large by the more modest standards of most other major volcanic provinces.

View southeast along a row of cinder cones aligned along the Great Rift. This old picture was taken before the modern roads and trails were built.

— U.S. Geological Survey photo by H.T. Stearns

The lava field erupted from the north end of the Great Rift, a zone of open fissures and volcanic vents that extends southeast almost 50 miles through several other lava fields of about the same age, almost to American Falls. The Great Rift follows the trend of Basin and Range faults, and surely relates to those rather than to the Yellowstone hotspot, which left this area in its wake several million years before these lava flows erupted.

In the Craters of the Moon area, as in the other lava fields along its length, the fissures of the Great Rift are filled with basalt and buried under lava flows and little cinder cone volcanoes. Nevertheless, the rift remains clearly visible in striking alignments of the basalt cinder cones along northwest trends. South of the Craters of the Moon lava field, four great sets of open fissures emerge from beneath the flows and strike southeastward across the volcanic plateau.

The Age of the Eruptions

Movie actresses and basalt lava fields often deceive us into thinking them much younger than they are. Basalt tends to look fresh for an amazingly long time because it offers so many porous opportunities for surface water to soak into the ground. So little water flows across basalt cinder cones and lava flows that they typically escape gullying for many thousands of years.

The first geologist to study Craters of the Moon, in 1902, concluded after very careful observation that the latest eruptions had happened within the last 150 years. The next geologist concluded, in 1928, that the latest eruptions probably happened long before that, at least several hundred years ago. Both were excellent geologists thoroughly

The lava level dropped after a basalt flow half buried and burned a tree, leaving this mold of the trunk. The hat on top gives an impression of tis size. —U.S. Geological Survey photo by H.T. Stearns

familiar with volcanic rocks who based their opinions on the general appearance of the area and the size of the biggest trees. Neither enjoyed the benefits of radiocarbon dating.

Modern radiocarbon dates on charcoal collected from beneath flows in the Craters of the Moon lava field show that activity there began about 15,000 years ago, and went through a minimum of eight periods of eruption that produced a total of at least 40 lava flows from 25 different vents. The periods of activity seem to have lasted less, perhaps much less, than a thousand years, and to have been separated by quiet periods that lasted from several hundred to as much as 2000 years. The most recent eruptions happened a bit more than 2000 years ago, so a new series of eruptions sometime soon would come right on schedule, if there is a schedule.

Cinder Cones

Most basalt eruptions begin as escaping steam blasts shreds of molten lava out of the vent in a noisily spectacular display of natural fireworks. By day, dark clouds of ash blow into the air and drift downwind in an ominous plume that darkens the sky and drops a steady fall of heavy ash. Meanwhile, the larger fragments land near the vent, each one adding its bit to the growing pile of basalt fragments, a cinder cone. At night, the red hot globs of molten lava shooting out of the crater trace their glowing trajectories across the dark sky, then roll down the slopes of the growing cinder cone, their glow fading as they cool.

The larger chunks of lava that litter the surfaces of most cinder cones are called bombs. Many are more or less streamlined because they were blown out of the vent as soft globs of molten lava, then aerodynamically shaped as they flew molten through the air. Some streamlined bombs taper at both ends into the shapes of weirdly twisted spindles. Bread crust bombs have a surface that really does suggest the crust on a loaf of French bread — if you can imagine a coal black loaf. The resemblance is no accident; bombs and loaves form their surfaces in exactly the same way as a hard crust forms, then breaks into patches as gas bubbles expand within.

Aa and Pahoehoe Flows

Some basalt flows, the kind called pahoehoe, have generally smooth surfaces that tend in places to become ropy and billowy. They look like black taffy. You can almost imagine riding a bicycle across the smoother parts of pahoehoe flows. Aa flows, on the other hand, are covered with a sea of angular blocks of rough basalt full of gas bubbles.

The ropy surface of a pahoehoe lava flow.

You can utterly ruin your boots in an hour or two of walking on such sharp rubble. It is hard to imagine more strikingly different surfaces, yet the rocks are essentially identical in all other respects. The most convincing evidence of that basic similarity is the occasional basalt lava flow that has a pahoehoe surface in one part, an aa surface in another.

Pahoehoe flows erupt with enough water in the lava to make it extremely fluid. As the upper surface of the flow begins to solidify, the lava within continues to move, rumpling the hardening surface in about the way hot cocoa rumples the thin skin on its surface. The interiors of pahoehoe flows are generally full of large gas cavities, and chunks of the rock feel light in the hand. As they blow off steam with increasing distance from the vent, such flows may develop an aa surface.

Aa flows contain just barely enough water that escaping steam fills the solidifying crust on the flow surface with gas bubbles as though it were the foaming head on a glass of beer. Continued movement of the molten lava within the flow breaks the bubbly surface into angular blocks full of the sharp edges of broken gas bubbles. Chunks of aa lava feel distinctly heftier than those broken from pahoehoe flows, and the rock is visibly less bubbly.

Lava Tubes

Many basalt flows are hollow inside because molten lava flowed out from beneath an already solid outer crust, leaving large parts of the flow an empty shell. The hollows commonly form long tubes that wind on and on, natural tunnels within the flow. Some of those lava tubes continue for miles.

Many lava tubes have long benches along their walls that mark a level where the flowing lava within stalled for a while, before it drained out. The walls and ceilings of many lava tubes are completely covered with black drips and dribbles of lava that make an elaborate ornamentation fully as exotic and almost as lavish as that in a limestone cavern.

A frozen lava cascade in a flow with a pahoehoe surface.
— U.S. Geological Survey
photo by H.T. Stearns

A sinkhole opened where the top of a lava tube collapsed near Indian Tunnel.

The roofs of underground hollows of any kind eventually fall, filling the hollow with broken rubble and generally opening a sinkhole on the surface. As lava tubes collapse, they first become a series of holes in the surface of the flow, eventually a long depression. Many collapsed lava tubes finally become stream valleys, the first stage in development of an erosional topography on the flow.

View inside Buffalo lava cave. The horizontal lines mark levels of the stream of lava. —U.S.Geological Survey photo by H.T. Stearns

292

Spatter Cones

Steam persistently blowing out of one place in a lava flow may cough out globs of molten basalt that accumulate around the vent to build a sort of free form chimney, a spatter cone. Most spatter cones are only a few feet high and as hollow inside as a chimney, but a few grow to the proportions of a small cinder cone. In many, the globs of basalt look almost like so many handfuls of black mud slapped carelessly onto each other to build the cone. You can see how the molten lava dripped and sagged before it finally solidified. Spatter cones generally mark steam vents on a lava flow, not the primary volcanic vent that produced the flow. Some lava flows have several spatter cones on them; many others have none.

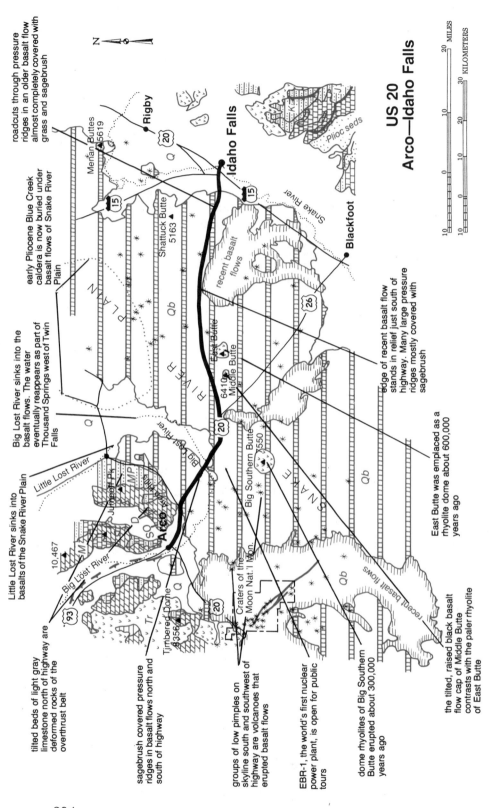

roadcuts through pressure ridges in an older basalt flow almost completely covered with grass and sagebrush

early Pliocene Blue Creek caldera is now buried under basalt flows of Snake River Plain

Big Lost River sinks into the basalt flows. The water eventually reappears as part of Thousand Springs west of Twin Falls

Little Lost River sinks into basalts of the Snake River Plain

tilted beds of light gray limestone north of highway are deformed rocks of the overthrust belt

sagebrush covered pressure ridges in basalt flows north and south of highway

groups of low pimples on skyline south and southwest of highway are volcanoes that erupted basalt flows

EBR-1, the world's first nuclear power plant, is open for public tours

dome rhyolites of Big Southern Butte erupted about 300,000 years ago

the tilted, raised black basalt flow cap of Middle Butte contrasts with the paler rhyolite of East Butte

East Butte was emplaced as a rhyolite dome about 600,000 years ago

edge of recent basalt flow stands in relief just south of highway. Many large pressure ridges mostly covered with sagebrush

Rigby

Idaho Falls

Blackfoot

Arco

Menan Buttes
5619

Shattuck Butte
5163

East Butte

Middle Butte
6410

Big Southern Butte
7550

Jumpoff Pk

10,467

Timbered Dome
8356

Craters of the Moon Nat.'l Mon.

Snake River

Big Lost River

Little Lost River

Big Lost River

SNAKE   RIVER   PLAIN

Plioc seds

recent basalt flows

recent basalt flows

Qb

Q

Tr

N

**US 20
Arco—Idaho Falls**

MILES

KILOMETERS

U.S. 20:
Arco — Idaho Falls
67 miles

Every high school class since 1931 has its inscription on the high cliffs that overlook Arco. They are Paleozoic limestone, the Surrett Canyon formation. The original sediments were laid down in shallow sea water during Mississippian time, about 300 million years ago. All the rocks along the road south of those cliffs are basalt and rhyolite, volcanic rocks erupted within the last million or so years.

Watch along the route across the Snake River Plain for the edges of basalt flows, low escarpments marked by abrupt changes in the plant cover — the fewer plants, the younger the flow. The low and inconspicuous mounds a mile or so across and several hundred feet high that rise here and there above the Snake River Plain are small shield volcanoes. They are the vents that erupted the basalt lava flows.

Isolated steep buttes that rise in sharp relief above the nearly level volcanic surface are younger rhyolite domes punched through the basalt. Rhyolite domes are so much steeper than basalt shield volcanoes because molten rhyolite is so much more viscous than molten basalt.

The Sink of the Big Lost River

The Big Lost River comes bustling down out of the mountains of central Idaho and through Arco full of purpose, as though it were a stream heading for some widely advertised destination. Then it gets onto the Snake River Plain, flows generally northeast, makes a dwindling marshy loop, and abandons all hope as it disappears into the porous basalt flows. The river begins to lose water to the basalt as soon as it crosses onto the Snake River Plain, and comes to its final end in a large expanse of seasonal marshland that it reaches only in flood years. Once underground, the water trickles south through the rubbly zones between lava flows. It eventually surfaces in the numerous large and small springs along the weeping north wall of the Snake River Canyon.

The Snake River Plain generally slopes gently southwards, largely because of the deposits of sediment the Big Lost River and other streams leave as they disappear into the basalt along its northern edge. However, the surface of the plain tends to be a bit higher in the center than at the edges, probably because that is where most of the eruptive centers are, the path of the hotspot. That slightly high center explains why the Big Lost River follows a path nearly parallel to the edge of the Snake River Plain. The marsh where it finally disappears is near the western margin of the Lemhi Valley, where active Basin and Range faults break the plain, opening fractures that help the basalt absorb water.

Big Southern Butte

The massive bulk of Big Southern Butte, the dominant landmark in this part of Idaho, looms in the distance south of the highway. It rises almost 2500 feet above the general level of the Snake River Plain, an island of rhyolite boldly standing watch over the sea of basalt. No one overlooks Big Southern Butte.

Big Southern Butte is an oversized rhyolite dome that erupted about 300,000 years ago. In fact, it may be two rhyolite domes: an older one in the more rugged southwestern part of the mountain, and a slightly younger mass that forms the smoother northeastern part.

The extremely viscous rhyolite magma contained so little steam that it erupted quietly — without steam, there is nothing to inspire volcanic commotion. The magma first intruded the older basalt flows, then broke through them to form the mountain we see. On its way up, the magma raised, gently tilted, and partially engulfed a large slab of older basalt flows to form the broad and gently sloping bench on the north side of the mountain. Younger basalt flows flooded around the base of Big Southern Butte after its coalesced rhyolite domes punched through the older flows.

Twin Buttes. Middle Butte with its flat summit of layered lava flows is in the foreground. —Bill Hackett photo

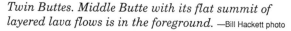

296

Twin Buttes

Twin Buttes, otherwise known as Middle and East Butte, also lie south of the highway. Although they look from a distance like small basalt cinder cones, both appear upon a closer look to be lesser versions of Big Southern Butte. All three buttes lie on a nearly straight line, quite possibly the trace of a Basin and Range fault.

East Butte is a small rhyolite dome that punched through the basalt lava flows some 600,000 years ago, according to age dates on the rhyolite. The talus slopes of broken rhyolite rubble on its flanks give East Butte the same general form as a cinder cone, which also consists of a rubble of loose chunks of rock.

Middle Butte is simply a raised and slightly tilted block of basalt — look at the layered rock below its squared off summit. Detailed studies of the earth's magnetic and gravity field around Middle Butte lead geologists to suspect that a mass of rhyolite must lie at shallow depth beneath it. Rhyolite is neither as magnetic nor as dense as basalt, so a mass of rhyolite at shallow depth causes the earth's magnetic and gravity fields to be slightly weaker than they would be if the rock beneath the surface were basalt. Intrusion of that rhyolite magma probably punched up the basalt flows above it like a piston to form Middle Butte. Without a sample of the buried rhyolite, it is impossible to get an age date.

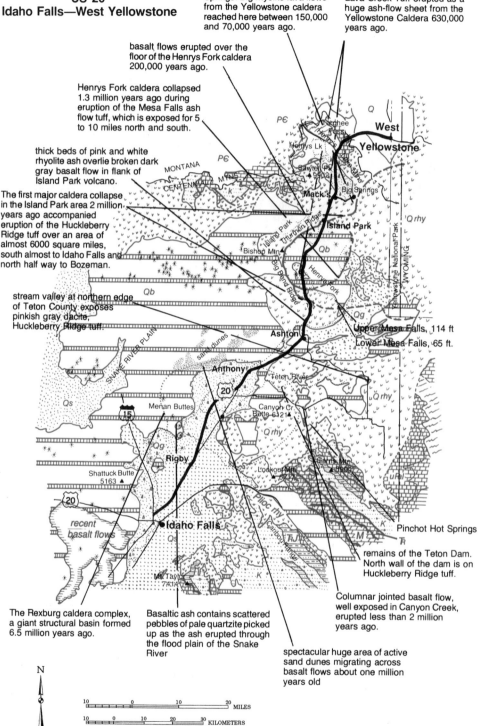

obsidian flow behind Big Springs. Big rhyolite lava flows from the Yellowstone caldera reached here between 150,000 and 70,000 years ago.

Lava Creek Tuff erupted as a huge ash-flow sheet from the Yellowstone Caldera 630,000 years ago.

basalt flows erupted over the floor of the Henrys Fork caldera 200,000 years ago.

Henrys Fork caldera collapsed 1.3 million years ago during eruption of the Mesa Falls ash flow tuff, which is exposed for 5 to 10 miles north and south.

thick beds of pink and white rhyolite ash overlie broken dark gray basalt flow in flank of Island Park volcano.

The first major caldera collapse in the Island Park area 2 million years ago accompanied eruption of the Huckleberry Ridge tuff over an area of almost 6000 square miles, south almost to Idaho Falls and north half way to Bozeman.

stream valley at northern edge of Teton County exposes pinkish gray dacite, Huckleberry Ridge tuff.

The Rexburg caldera complex, a giant structural basin formed 6.5 million years ago.

Basaltic ash contains scattered pebbles of pale quartzite picked up as the ash erupted through the flood plain of the Snake River

Columnar jointed basalt flow, well exposed in Canyon Creek, erupted less than 2 million years ago.

spectacular huge area of active sand dunes migrating across basalt flows about one million years old

remains of the Teton Dam. North wall of the dam is on Huckleberry Ridge tuff.

Pinchot Hot Springs

Upper Mesa Falls, 114 ft
Lower Mesa Falls, 65 ft

West Yellowstone

Big Springs

Island Park

Mack's

Ashton

Anthony

Rigby

Idaho Falls

Menan Buttes

Shattuck Butte 5163 ▲

Mt Taylor 7414

Bishop Mtn

Lookout Mtn

Garns Mtn

Canyon Cr Butte 6121 ▲

Teton River

Henrys Lk

Sawtell Pk 9902

Targhee Pass

MONTANA

CENTENNIAL MTNS

SNAKE RIVER PLAIN

recent basalt flows

sand dunes

WYOMING

Yellowstone National Park

N

10   0   10   20 MILES
10   0   10   20   30 KILOMETERS

U.S. 20:
Idaho Falls — Targhee Pass
95 miles

The highway cuts across a broad loop of the Snake River for 18 miles north of Idaho Falls, then crosses the river about midway between Rigby and Rexburg. The route north of St. Anthony follows the Henry's Fork River along almost its entire length to its headwaters in the mountains near Targhee Pass.

Except in the Centennial Mountains near Targhee Pass, all the bedrock along the way is basalt and rhyolite of the Snake River Plain. Most of the rocks at the surface are basalt, almost certainly a thin veneer covering a much larger, but mostly unseen, volume of rhyolite beneath. Although the eastern end of the Snake River Plain has seen most of the more recent volcanic activity, very little fresh basalt appears along this highway. A deeply obscuring mantle of river gravel, soil, and vegetation covers most of the volcanic rocks.

Near Targhee Pass, the highway crosses tightly folded Paleozoic sedimentary formations that belong to the northern Rocky Mountains. They are in the eastern end of the Centennial Range, which trends nearly east to west, instead of from north to south like most other ranges in the region. Basin and Range faults have not yet chopped it up.

Section drawn across the line of Highway 20 about five miles south of Rexburg. The gently dipping faults formed with the northern Rocky Mountains. The steeply dipping faults belong to the Basin and Range.

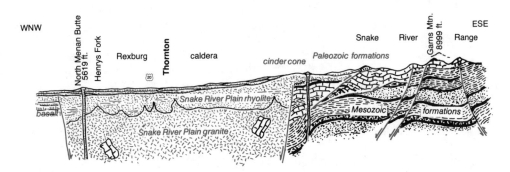

The Rexburg Caldera

Between Rigby and Rexburg, the highway crosses the Rexburg caldera, the faint scar of an enormous rhyolite volcano, a resurgent caldera that was active about 6.5 million years ago. Later volcanic eruptions so thoroughly obliterated the outlines of the old caldera that the volcano reveals absolutely nothing of itself to people travelling along the road. It appears only through study of geologic maps, and even with that broad perspective you need both a bit of conjecture and the knowing eye of faith to see much of the volcano.

Our geologic section shows granite at shallow depth beneath the Rexburg caldera. That is at least slightly conjectural; no Pliocene granite is exposed anywhere in the Snake River Plain. Nevertheless, we think it must exist, that large volumes of rhyolite magma must have crystallized into granite without erupting.

What will appear in the Snake River Plain after millions of years of erosion have stripped the basalt and rhyolite off its surface? To phrase the same question in other words, what would a deep hole drilled next year reveal? We suggest that the rocks at depth are mostly granite, a long and broad northeast trend of granite, cut by hundreds or thousands of basalt dikes that trend mostly northwest. The granite is the wake of the Yellowstone hotspot; the basalt dikes are the feeders for the younger lava flows associated with Basin and Range faulting.

North of Ashton, the road crosses the Henry's Fork caldera, which is clearly visible from the road, obvious on the geologic map, and not in the least conjectural.

The Henry's Fork and Island Park Calderas

Big Bend Ridge, which the highway crosses just north of Ashton, got its name from its map outline. It swings west through a large and nearly semi-circular arc that brings it around to the area near Island Park, where the highway crosses it again. That horseshoe ridge of rhyolite encloses the Henry's Fork caldera, a nearly circular basin about 13 miles across. Big Bend Ridge is the ruin of a large rhyolite volcano, a resurgent caldera similar to the one now active in Yellowstone Park. That old volcano collapsed during a final catastrophic eruption of approximately 1.3 million years ago. Younger volcanic rocks erupted from the Yellowstone volcano bury the eastern end of Big Bend Ridge and the Henry's Fork caldera.

Watch for road cuts in pale gray rhyolite where the highway crosses Big Bend Ridge north of Ashton. On sunny days, flat surfaces on large crystals of sanidine feldspar glitter as though they were thousands of little mirrors pasted onto the road cuts. Many of the road cuts also expose beautiful pumice, a glistening foam of white volcanic glass with blocky little crystals of milky sanidine and dark crystals of quartz embedded in it. Some of the pumice was so twisted while it was molten and erupting that it looks like pulled taffy. Some of it is so light and foamy that it will float on water.

The volumes of large rhyolite eruptions from resurgent calderas would be truly incredible if they were not so clearly recorded. Some two million years ago, for example, the Henry's Fork volcano covered almost 6000 square miles of Idaho and Montana with red hot ash, the Huckleberry Ridge tuff. Not one living thing in all that area, as well as a considerable band of surrounding countryside could possibly have survived the holocaust. The side effects of atmospheric pollution and general climatic cooling must have been awful.

The last 200 years have brought several rhyolite eruptions around the world that were certainly major in human terms, although extremely minor by the atrocious standards of resurgent calderas. Each brought about great human suffering by filling the upper atmosphere with a haze of fine ash, a "dry fog," that dimmed the sun for months, causing abnormally cold weather and widespread crop failures. An Indonesian eruption in 1815, for example, caused the "year without a summer" that brought widespread famine to the New England states. Imagine what a major eruption of a resurgent caldera would do. Surely those monsters must fill the sky with ash for months, quite possibly for years, causing a long volcanic winter and widespread famine. We can all consider ourselves fortunate that the world contains only a few active resurgent calderas, and that they erupt so infrequently.

View north over Menan Buttes. The Snake River flows past the southern edge of the buttes. —Bill Hackett photo

The Menan Buttes

The two large Menan Buttes, a few miles west of the road and about 15 miles north of Idaho Falls, are basalt volcanoes that stand on the floodplain of the Snake River, near the edge of the Rexburg caldera. The largest is about 800 feet high, and has a crater 300 feet deep in its summit. Several smaller, but otherwise fairly similar, volcanoes make the rest of the cluster. The easiest way to reach the Menan Buttes is to drive south from Highway 33 on Road 400W, which follows the west side of the buttes. Various side roads lead directly onto them.

The Menan Buttes consist of a wild mixture of basalt cinders and stream rounded pebbles, all stacked in steeply sloping layers. Many of the pebbles are broken. It seems that basalt magma rising into the water saturated floodplain gravels generated a long series of steam explosions violent enough to shatter the pebbles. Imagine the scene:

Volcanoes like the Menan Buttes form where rising magma boils groundwater, flashing it into steam which blows ash out of the vent, along with pebbles from the river gravels caught up in the eruption.

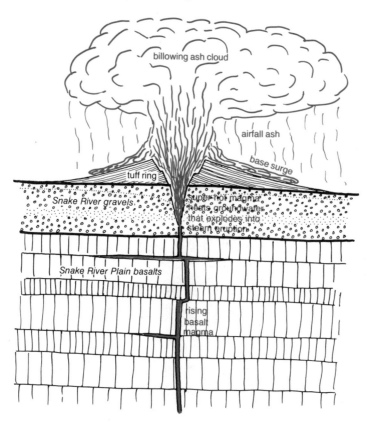

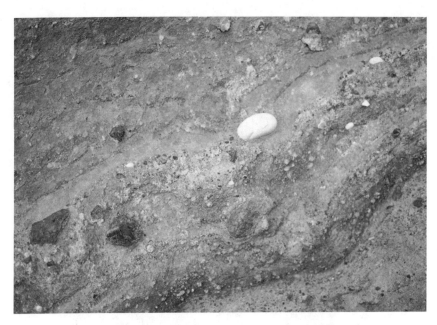

Rounded pebbles of white quartzite in the dark basalt

Each mighty blast of steam blew a dark cloud of basalt cinders mixed with pebbles into the air. Each successive cloud of debris settled to add a new layer to the slope of the rapidly growing volcano.

The layers stand at an angle of about 20 degrees and thicken down the slope. Excavations into the buttes show the largest chunks of debris concentrated at the base of each layer, with the grain size diminishing more or less regularly upward. Evidently, the explosive blasts of steam shot the smaller chunks higher, so they came down last.

Maps clearly show each of the Menan Buttes tailing out northeast of the crater into an elliptical outline. That northeast elongation makes it seem likely that the eruptions happened during the winter, when the prevailing winds in this region blow from the southwest. Imagine those tremendous clouds of steam hanging in the icy winter air, looking as solid as though they were made of cast iron.

Their position on the floodplain of the Snake River makes it clear that the Menan Buttes can't possibly be very old, as such things go. They are much too young to qualify as a late stage of activity in the Rexburg caldera. We think it is more reasonable to include them among the many volcanic centers associated with Basin and Range faulting.

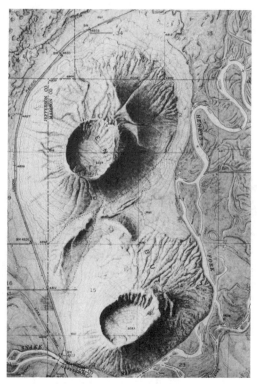

The elliptical outlines of the Menan Buttes record the direction the wind blew during the eruptions. —U.S. Geological Survey Menan Buttes Quadrangle

Big Spring

Fishing is strictly against the rules at the bridge over the spring run, where schools of oversized, overfed and perpetually ravenous trout beg tirelessly for handouts. Big Spring is one of the 40 largest springs in the country; it produces about 120,000,000 gallons of water each day, enough to supply a city of nearly 1,000,000 people. It is the main source of the Henry's Fork River. All that water falls as rain in the wet high country of the Yellowstone Plateau, soaks into the ground, then percolates through the volcanic rocks to Big Spring. It emerges in all seasons at a steady temperature of 52 degrees Fahrenheit, just slightly above the mean annual temperature of the area. Most springs produce water at the local mean annual temperature.

Big Spring issues from the base of a large obsidian lava flow on the western flank of the Yellowstone Plateau. The brown and black pebbles in the bed of the spring run are obsidian, volcanic glass. It is a rare and interesting kind of rock.

Some Thoughts about Obsidian

Obsidian has essentially the same chemical composition as ordinary rhyolite or granite, but differs from them in having hardened into glass as it cooled, instead of crystallizing into an ordinary rock. Think of it as solidified magma. For many years, geologists attempted to explain the glassiness of obsidian by arguing that it cooled too quickly to allow time for crystals to form. But that interpretation breaks down when you consider that obsidian flows are typically more than one hundred feet thick; many are several hundreds of feet thick. It would be physically impossible to chill such a large mass of lava quickly. The solution to the problem probably lies instead in the one small difference between the chemical compositions of obsidian and rhyolite.

Unlike rhyolite and most other rocks, obsidian contains virtually no water. The lack of water makes the magma even more extremely viscous than it would otherwise be, thus explaining the great thickness of obsidian flows — a more fluid magma would run out into thinner flows. At the molecular level, high viscosity translates into low mobility; the atoms are not free to move about within the magma as they must if they are to assemble themselves into crystals. So, the magma cools into a glass.

Their lack of water also causes obsidian flows to erupt quietly, without the usual volcanic fireworks displays that depend entirely upon steam for their inspiration. If the same magma had picked up some water at shallow depth, it might very well have exploded into a rhyolite ash flow. If it had absorbed water at greater depth, it would have crystallized into granite before it could reach the surface. So much hinges upon the chance encounters of water and magma.

Water and magma are now setting the stage beneath Yellowstone Park for one of those scenarios, but it is impossible to know which. Is the magma there absorbing water while it is still deep enough to crystallize quietly into solid granite before it can erupt? Or is the situation hopeless, the magma already so close to the surface that infiltrating water is charging it with steam that will someday explode in a great eruption? Or perhaps the water and magma beneath the park are somehow missing contact, so the magma will eventually erupt quietly, as obsidian. One of these scenarios will determine the future of the volcano, as well as of the national park and everything associated with it. There is no way to predict the locations and timing of such eruptions but past experience suggests that most of the rhyolite will erupt as giant ash flows.

The shifting sand dunes north of St. Anthony — they should be a state park.

The Saint Anthony Dune Field

The biggest tract of sand dunes in Idaho covers an area of approximately 175 square miles, which reach to within four miles of St. Anthony. To drive well into the active dunes, follow the paved road west from St. Anthony five miles to the road a mile west of Parker, then turn north and drive another three miles. The active dune field trends generally northeast about 35 miles, and is five miles wide. Older dunes now stabilized under grass cover much of the surrounding area. They must record a period when the climate was drier than it now is.

Most of the dunes curve into sharp crescents that end in a pair of horns. That is the characteristic shape of dunes that form in places where sand is abundant and the winds strong enough to move it blow consistently from a single direction. The horns of the crescents and the steep slopes of the dunes consistently face northeast, the direction of sand movement. Evidently, the stiff southwesterly winds of winter drive the sand.

From a distance, the dunes are tan. A good magnifying glass reveals that the sand consists mostly of white quartz slightly peppered with occasional grains of black basalt. A thin stain of iron oxide

Sand drifted into the openings in the crest of this pressure ridge in a basalt flow north of St. Anthony.

on the quartz grains gives the dunes their tan color. Quartz does not exist in the basalts that cover most of the Snake River Plain. Evidently, the Snake River and its tributaries bring the quartz in from the mountains, then the wind whips it off the floodplain and into the dunes.

Sand dunes boldly defy our preconceived expectations. Ask almost anyone how the wind will dispose of a lot of loose sand, and the reply will almost certainly be that the wind will scatter it all over the place. But in fact, you can easily see that the wind sweeps sand up into neat piles, dunes with their keenly swooping sculptured forms. That happens because a pile of sand is soft.

In fact, softness is the very essence of sand dunes, their reason for existing. Imagine yourself and a friend batting marbles back and forth across a terrazzo floor. Now imagine what would happen if someone were to lay a scrap of soft carpet somewhere on the floor. No problem. All the marbles would soon be on the carpet because its softness would prevent their bouncing off. Sand dunes work the same way. They grow larger as sand grains land on their soft surfaces, and don't bounce off. Bare basalt surfaces between the dunes correspond to the hard terrazzo floor — the wind whips sand across them and into the dunes.

The Grand Tetons, view from north of Driggs.

The Grand Tetons

Throughout the country east of Ashton and Rexburg, the three jagged peaks of the Grand Teton Range of Wyoming scrawl their distinctive profile along the eastern horizon. The view from the west of a backdrop of mountains rising beyond a neatly domesticated foreground of cultivated farmland is utterly different from the wilder eastern aspect of the range familiar on so many calendars and travel posters.

The Grand Tetons are a slice of basement rock hoisted many thousands of feet along a fault that defines the eastern front of the range, also the eastern margin in this area of the Basin and Range. Project the trend of the Tetons south, and it lines up with that of the Wasatch Front in Utah. The same trend projected north passes through the Yellowstone volcano and beyond into the steep western face of the Madison Range, in Montana.

The steep mountain wall of the famous eastern face of the Teton Range is the fault scarp, which appears to have risen within the last few million years, and probably continues to rise. Ice age glaciers carved the ancient basement rocks that make those peaks into the snaggled row of sharp teeth we see today.

All That Remains of the Teton Dam

Many people regard the remains of the Teton Dam as the most bizarre spectacle in Idaho. It is certainly worth seeing. A sign beside Highway 33 about 10 miles east of Sugar City points the way: Drive one mile north of the road, park in a paved lot rapidly going to grass, and look over the edge.

The Teton Dam was a Bureau of Reclamation project that was supposed to provide construction jobs, flood control, irrigation water, electricity, and recreation — something for almost everyone. Construction began in 1972, after almost 20 years of haggling debate over various questions of water rights, cost effectiveness, need for irrigation water and electricity, and preservation of natural values. Actually, the basic question was whether the dam would materially benefit anyone beyond the Bureau of Reclamation and its contractors.

About breakfast time on June 5, 1976, just as the reservoir was nearly filled for the first time, the dam began to leak in several places. Bulldozers dispatched to plug leaks were promptly lost, vanished into the growing wash of debris — the situation had already deteriorated far beyond the finger in the dike stage. The leaks grew rapidly until just before noon, when the dam abruptly washed out. The failure released some 80 billion gallons of water in a sudden flood that devastated Sugar City an hour later. The flood went on down through Rexburg and Idaho Falls, and into American Falls Reservoir, which absorbed it. Look for the high water mark on the flood-scoured canyon walls below the remains of the dam.

Poor foundation rock and improper design caused the disaster. The dam was poorly located and improperly designed for the location. Water seeping through fractures in the basalt under the dam eroded a tunnel through the silty clay used in the lower part of the structure. Then the undermined dam collapsed. The design should have compensated in some way for the high permeability of the underlying bedrock.

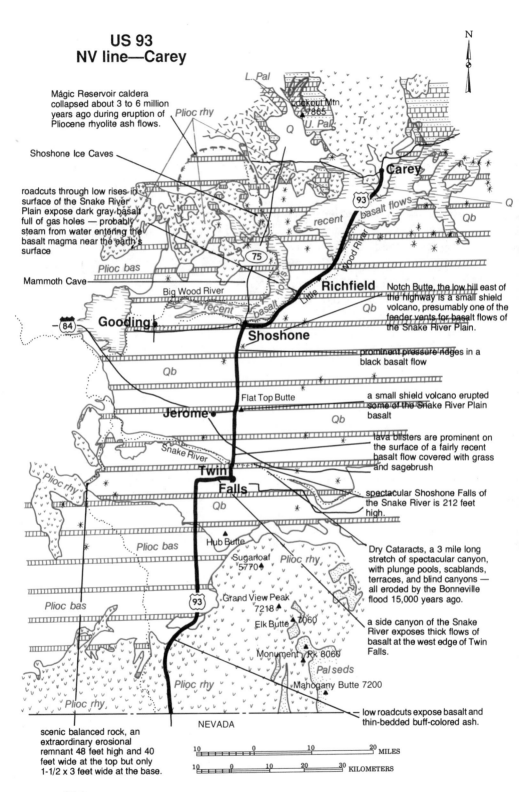

N

Mágic Reservoir caldera collapsed about 3 to 6 million years ago during eruption of Pliocene rhyolite ash flows.

Plioc rhy

Shoshone Ice Caves

roadcuts through low rises in surface of the Snake River Plain expose dark gray basalt full of gas holes — probably steam from water entering the basalt magma near the earth's surface

L. Pal

Lookout Mtn. ▲7865

U. Pal.

Q

Tr

Carey

93

recent

basalt flows

Wood River

Qb

Q

Plioc bas

75

Mammoth Cave

Big Wood River

recent

Lime

Richfield

basalt flows

Notch Butte, the low hill east of the highway is a small shield volcano, presumably one of the feeder vents for basalt flows of the Snake River Plain.

84

Gooding

Qb

Shoshone

prominent pressure ridges in a black basalt flow

Qb

Flat Top Butte

a small shield volcano erupted some of the Snake River Plain basalt

Jerome

Qb

lava blisters are prominent on the surface of a fairly recent basalt flow covered with grass and sagebrush

Snake River

Twin Falls

Qb

spectacular Shoshone Falls of the Snake River is 212 feet high.

Dry Cataracts, a 3 mile long stretch of spectacular canyon, with plunge pools, scablands, terraces, and blind canyons — all eroded by the Bonneville flood 15,000 years ago.

Plioc bas

Hub Butte

Sugarloaf 5770▲

Plioc rhy

Plioc bas

93

Grand View Peak 7218▲

Elk Butte ▲7060

a side canyon of the Snake River exposes thick flows of basalt at the west edge of Twin Falls.

Monument Rk 8060▲

Pal seds

Plioc rhy

Mahogany Butte 7200

low roadcuts expose basalt and thin-bedded buff-colored ash.

Plioc rhy

NEVADA

scenic balanced rock, an extraordinary erosional remnant 48 feet high and 40 feet wide at the top but only 1-1/2 x 3 feet wide at the base.

10   0   10   20 MILES

10   0   10   20   30 KILOMETERS

The flood scoured Snake River Canyon from the highway bridge at Twin Falls. The Bonneville flood filled this canyon to the brim with muddy water.

U.S. 93:
Nevada Line — Carey
102 miles

All the rocks exposed along Highway 93 between the Nevada line and Carey are volcanic, all part of the Snake River Plain. You see some rhyolite near the south end of the route, only basalt elsewhere.

As everywhere on the Snake River Plain, you can estimate the relative ages of basalt flows by comparing their soil and plant cover. The youngest are still naked — black expanses of jagged basalt. Others are so thoroughly blanketed that no bedrock is exposed. Changes in soil and plant cover so sharply mark the boundaries between the younger flows that geologists have been able to map most of them, and give them individual names.

The Northern Rocky Mountains

The Northern Rocky Mountains rise on the horizon north of Carey. Peaks in the western part of the view consist mostly of granite, part of the southern Idaho batholith; those farther east are tightly folded Paleozoic sedimentary formations. Large areas of both the granite and the folded sedimentary rocks lie beneath deep blankets of rhyolite ash — two generations of rhyolite.

Most of that ash is Challis volcanic rock that erupted in central Idaho during Eocene time, approximately 50 million years ago. That was long before volcanic activity began in the Snake River Plain. Some of the rhyolite northwest of Carey erupted during Miocene time, a date that suggests some relationship to the volcanic activity that built the Columbia Plateau. More Miocene rhyolite covers much of the Cassia Mountains near the southern end of the route.

The Cassia Mountains

The Cassia Mountains lie low on the horizon east of the highway, about midway between the Nevada line and Twin Falls. Like the other ranges in this part of Idaho, they are a block sliced out of the Rocky Mountains along the much younger faults of the Basin and Range province. Those blocks rise or fall, tilt, and pull away from each other as movements in the mantle stretch the earth's crust.

Rocks in the Cassia Range include a core of sedimentary formations that were deposited during Paleozoic time, then lay undisturbed until they were folded as the Rocky Mountains formed during late Mesozoic

Section drawn along a line just east of Highway 93 showing

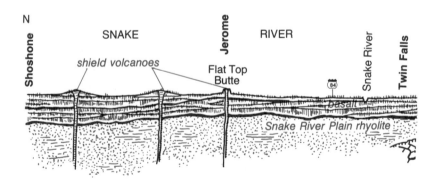

time, sometime around 70 to 80 million years ago. A blanket of much younger volcanic rocks, mostly rhyolite ash flows that belong to the Snake River Plain, covers most of the older folded rocks.

Rhyolite

Through the 15 miles just north of the Nevada line, the highway crosses rhyolite that belongs to the Snake River Plain. Most is ash flows that were still hot enough when they finally settled to weld themselves into more or less solid rock, welded ash. They vary in color from white through shades of red to almost black.

We normally think of rhyolite as a pale rock, so it seems a bit strange to see it in shades of dark red and black. Those darker colors commonly exist in rhyolites that consist largely of glass, instead of minute crystals. The internal molecular structure of glass is that of a congealed liquid, rather than of a crystalline solid. A small amount of dissolved iron colors glass in the same way that a small drop of ink stains a large amount of water. In crystalline rhyolite, iron generally forms minute grains of black magnetite that lightly pepper the rock, but do not stain it.

Given time, glassy rhyolites invariably crystallize into ordinary rhyolite, losing their dark color as the iron segregates into tiny grains of magnetite. You rarely see glassy rhyolite that is more than 20 or 30 million years old; never see it in volcanic piles much more than about 50 million years old.

the Cassia Mountains rising south of the Snake River Plain.

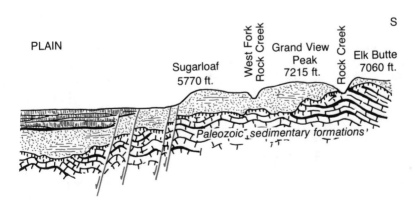

A rhyolite ash flow exposed along the road about five miles north of the state line contains Thunder Eggs, colorful concoctions of chalcedony and quartz crystals about the size of eggs. They make beautiful polished slabs. A few rhyolite ash flows contain such lumps, although most do not. We have never seen a convincing explanation of how they form.

Chalcedony, often called agate, is another variety of quartz, the mineral form of silica, silicon dioxide. Silica, is the most abundant chemical constituent of rhyolites, so it is not difficult to imagine why agate and quartz crystals might fill cavities in a rhyolite ash flow. The problem arises in imagining how the cavities formed. What could open hundreds of cavities the size of eggs inside a rhyolite ash flow?

If you are lucky enough to saw a Thunder Egg along just the right direction, the cut will show a curious round structure about the size of a marble on one side of the filled central cavity, a corresponding round hole, also filled, on the opposite side. The marble would fit into the hole if the central cavity could be emptied and the Thunder Egg collapsed. Evidently, the Thunder Egg began as a round object about the size of a marble that for some reason expanded to form a much larger cavity, which then filled with quartz or chalcedony.

Basalt

Except for the rhyolites near the Nevada border, all the rocks along the road are basalt lava flows of various ages that become generally younger northward. Most of the route between Shoshone and Carey crosses basalt flows so very young that they have only a sparse covering of soil and plants. Such young lava flows so far west on the Snake River Plain could hardly have anything to do with the Yellowstone hotspot, which left this area behind about ten million years ago. The young flows are almost certainly associated with the Basin and Range faults now beginning to crack the Snake River Plain into blocks.

Flat Top Butte, the hill with the transmitting antennas on it directly east of Jerome, and Notch Butte, east of the road a few miles south of Shoshone, are small shield volcanoes that erupted some of the younger flows. The road passes within sight of a number of other small shield volcanoes, but most of them are hard to recognize. Watch for inconspicuous low mounds a mile or so in diameter that rise a few hundred feet above the general level of the Snake River Plain. Small shield volcanoes are typically so low and flat, such inconspicuous blisters, that they are hard to see.

A pressure ridge makes a low rise in the Snake River Plain, beside Highway 93 a few miles north of Twin Falls.

In several places, especially between Twin Falls and the junction with Interstate 84, you can see excellent pressure ridges from the road. Watch for low mounds and ridges of black rock, a hundred feet or so across, that rise a few feet above the sagebrush.

Pressure ridges form as the solid crust developing on the cooling flow rafts along on the still moving lava beneath. Where the flow crosses rough ground, pieces of that solid crust jam together like crumpling slabs of ice in a winter river. The deep crack that follows the crest of each pressure ridge formed as the crumpling slab broke. In a few places oozes of molten basalt squeezed like putty through those cracks, making globs that look like black toothpaste squashed from a giant tube.

One flow just north and northeast of Shoshone, so recent that it is still almost bare rock, fills an old valley, presumably the former valley of the Wood River. The present Big Wood River follows a topographic low along the northwest edge of that flow, while the Little Wood River follows a similar depression along its southern margin. They join at Gooding, where the flow ends, to follow the original valley.

A Pair of Waterfalls

Shoshone Falls, five miles east of Twin Falls, is often called the Niagara of the West. That is no idle exaggeration. Shoshone Falls are 200 feet high, about 40 feet higher than Niagara, nearly 1000 feet wide, and discharge a comparable volume of water, or would if so much were not diverted to irrigation. Like Niagara Falls, both Shoshone

Shoshone Falls are still magnificent despite considerable diversion of water. The falls pour over a ledge of hard rhyolite.

Falls and Twin Falls, three miles farther east, are much exploited for electric power generation. Despite the defacements, more than enough natural beauty remains at both falls to richly reward a visit.

At the falls, the Snake River has cut through the black basalt lava flows of the Snake River Plain, and is now sawing ever so slowly into the pale rhyolite beneath. Those exposures helped convince geologists that rhyolite generally lies just beneath the basalt you see nearly everywhere on the surface of the Snake River Plain.

Shoshone and Twin Falls both plunge into basins eroded during the Bonneville flood, rather than into plunge pools of their own making. Large whirlpool eddies swirling in the dark floodwater that filled the canyon to its brim plucked blocks from the bedrock to carve deep holes in the canyon floor. After the flood passed, the river plunged over the edges of the basins in broad waterfalls. The 15,000 or so years of normal stream flow since the Bonneville flood passed have brought almost no further erosion of the rhyolite ledges at the lips of the falls, doubtless because the river is clear most of the time.

Clear running water can no more erode hard rock than a summer breeze can clean old paint and grime off a building. Running water, like blowing air, erodes only when it carries particles of abrasive, such as sand. In both cases, the process is basically a matter of sandblasting, the only difference being whether water or air carry the abrasive. Muddy streams quickly carve the lips of waterfalls, clear streams do not. That is why you see waterfalls only on clear streams.

The Snake River Canyon also contains a number of features that look like they might be dry cataracts, but probably are not. They

Twin Falls as they looked in 1902.
—U.S. Geological Survey photo by H.T. Stearns

include blind canyons, alcoves recessed into the valley walls, and holes that resemble abandoned plunge pools. Like the two big waterfalls, these probably formed where giant eddies in the Bonneville flood scoured the canyon walls and floor.

In most ways, basalt is an extremely hard and durable rock. But the vertical columns you see all along the canyon walls make the rock peculiarly vulnerable to catastrophic erosion beneath great floods. The columns easily separate into chunks of rock small enough to roll beneath the torrent.

Ice Caves

The Shoshone Ice Caves are the most accessible of many that exist in the Snake River Plain. Ice forms in caves that air can enter only by sinking vertically, only when the air outside is colder and therefore denser than that inside. Ice does not form if avenues exist that permit warm summer air to circulate through the cave. People who develop an ice cave to admit visitors must be very careful to preserve the natural pattern of air flow, lest they melt the ice.

Caves in basalt flows are especially likely to fill with ice because they generally are lava tubes that form as still molten basalt runs out from under a solid crust. Then the cave opens vertically to the surface as the roof begins to collapse. Meanwhile, debris dumped into the lava tube as the roof falls elsewhere tends to block horizontal air circulation.

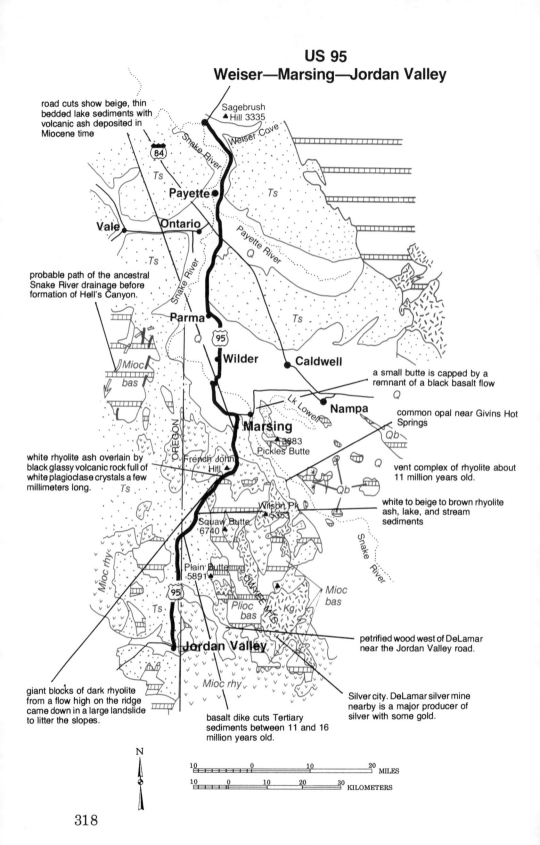

US 95
Weiser—Marsing—Jordan Valley

road cuts show beige, thin bedded lake sediments with volcanic ash deposited in Miocene time

Sagebrush
▲ Hill 3335

Weiser Cove

Snake River

84

Ts

Payette ●

Vale ●   ○ **Ontario**

Payette River

Q

Ts

Snake River

probable path of the ancestral Snake River drainage before formation of Hell's Canyon.

Parma ●

Q

95

Ts

Wilder ●   ● **Caldwell**

a small butte is capped by a remnant of a black basalt flow

Q

Lk Lowell

Mioc bas

Marsing ●   ● **Nampa**

common opal near Givins Hot Springs

Qb

▲ 3883
Pickles Butte

OREGON

French John Hill ▲

white rhyolite ash overlain by black glassy volcanic rock full of white plagioclase crystals a few millimeters long.   Ts

Q

vent complex of rhyolite about 11 million years old.

Qb

white to beige to brown rhyolite ash, lake, and stream sediments

Wilson Pk
▲ 5353

Squaw Butte
▲ 6740

Snake River

Mioc rhy

Plain Butte
▲ 5891

95

Ts

SILVER MTS

Plioc bas

Kg

Mioc bas

petrified wood west of DeLamar near the Jordan Valley road.

Jordan Valley ●

Mioc rhy

giant blocks of dark rhyolite from a flow high on the ridge came down in a large landslide to litter the slopes.

basalt dike cuts Tertiary sediments between 11 and 16 million years old.

Silver city. DeLamar silver mine nearby is a major producer of silver with some gold.

N

```
10        0        10        20  MILES
10     0     10     20     30  KILOMETERS
```

U.S. 95:
Weiser — Jordan Valley, Oregon
112 miles

The road follows the valleys of the Payette and Snake Rivers all the way between Weiser and the area near Marsing. You don't see much rock along that road. Very young river sediments floor the valleys, completely hiding the Columbia Plateau basalt beneath. A few exposures of soft sediments deposited in lakes impounded behind those flows exist east of the road in the area north of Perma.

Virtually all the hard bedrock exposed along the road is south of Marsing. It is all volcanic. Other rocks are soft sedimentary deposits, sand, silt, and mud. Part of that stuff accumulated in lakes impounded behind the lava flows; the rest is desert sediment laid down during the millions of dry years of late Miocene and Pliocene time. That interval started sometime after things calmed downed in the Columbia Plateau, perhaps about 13 or 14 million years ago, and continued until the beginning of Pleistocene time, about three million years ago.

Most of the volcanic rocks along the southern part of the route erupted during late Miocene time as the lava lake in southeastern Oregon slowly crystallized. This southern part of the Columbia Plateau includes both basalt and rhyolite. As everywhere, the basalt is easy to recognize by its very dark color and persistent tendency to fracture into vertical columns. Recognize the rhyolite by its light color, generally a gray so pale that it borders on white.

A remnant of a basalt flow caps a small hill north of Marsing.

Younger volcanic rocks erupted during Pliocene and Pleistocene time, almost certainly in association with Basin and Range faulting. Those flows appear mostly in scattered buttes well east of the highway, but the hill just north of Marsing is close enough to see; its flat cap is a Pliocene basalt flow. Jordan Craters lava field, near Jordan Valley, erupted recently enough to look fairly fresh. That lava field is at the southeastern end of the Brothers fault zone, which cuts northwestward across eastern Oregon, and marks the northern limit of Basin and Range faulting.

The Owyhee Mountains

The Owyhee Mountains have had a difficult past; they were too close for comfort to the big impact in southeastern Oregon. The older rocks there are granites that look about like those in the Idaho batholith of central Idaho, and are the same age, late Cretaceous. They almost certainly are an outpost of the Idaho batholith, now isolated from the main mass by the broad swathe of the Snake River Plain.

Enormous volumes of sparkling white Miocene rhyolite ash exploded across the Owyhee Mountains; it probably came from both the crystallizing lava lake in southeastern Oregon and from the first stages in the development of the Snake River Plain. Other masses of rhyolite magma, those with less steam in them, rose quietly to form bulging rhyolite domes and lava flows. Meanwhile, basalt lava flows of the Columbia Plateau flooded valleys in the Owyhee Mountains leaving the higher peaks standing free.

The troubles of the Owyhee Mountains have not ended. Basin and Range faults are now slicing the region into blocks and pulling them away from each other, converting these already isolated mountains into even more isolated fault block ranges.

Petrified Wood

Much of the volcanic ash in southwestern Idaho contains abundant petrified wood. It provides nice souvenirs of the forests that evidently covered the region in part of late Miocene time, as well as further evidence that a warmly humid climate prevailed then.

Rainwater saturated the ash, dissolved part of its content of silica, and redeposited it in the buried wood, petrifying it. The process depends upon the tendency of water soaked into volcanic ash to become slightly alkaline, thus capable of dissolving silica. Meanwhile, the decomposing wood becomes slightly acidic for basically the same reason that fermenting sauerkraut and dill pickles become mildly acidic. Silica is insoluble in acidic environments, so it precipitates in the slightly acidic wood to form the silica minerals opal or chalcedony.

It cannot be true, as we often read, that the depositing silica replaces the wood, molecule for molecule. The problem is that molecules in wood are not the same size as those of silica, so simple replacement would cause the wood to change its size and shape — you would distort a building if you were to replace its bricks with others of different sizes. Petrified wood is not distorted; it looks just like the original. In any case, woody tissues still exist within petrified wood; you can easily see them through a microscope. The depositing silica simply saturates the open pores in the wood in much the same way that oil sealer saturates a board.

Neither is it necessarily true, as we also read, that millions of years must pass before wood can petrify. No one actually knows how long the process takes, and there is no reason to assume that it always goes at the same rate. Petrifaction can happen very quickly; some of the dead trees around the expanding hot spring areas in Yellowstone Park, for example, are partially petrified where they stand. Most petrified wood is still as plump and round as it was in life, not at all squashed under the weight of the ash that buried it. That plumpness certainly suggests rather rapid petrifaction, long before the ash had time to settle and compact.

Growth rings preserved in the petrified wood of southwestern Idaho may have something to tell us. The warm and wet climate of late Miocene time appears to have been one aspect of a period of drastic climatic aberration that may well have involved the entire earth. If the cause involved some change in the behavior of the Sun, that may well appear in a departure from the 11-year sunspot cycle we see in the growth rings of modern trees. On the other hand, if the meteorite impact responsible for the Columbia Plateau also affected the climate, then the sunspot cycles in the wood should show no change.

Valley-Fill Sediments

Long segments of the route cross areas covered by younger sedimentary rocks: unconsolidated basin-fill deposits, wind blown dust, lake deposits, stream sediments, even lignite coal. In many areas, this material is still accumulating in the Basin and Range valleys as they drop.

The basin-fill deposits accumulated mainly because the climate has been extremely dry in this part of Idaho ever since the wet period of late Miocene time ended. Soil erosion is very rapid in dry regions, where raindrops splash directly onto the barren soil. Stream flow in such dry regions is generally too meager to carry all the sediment, so it accumulates on the continent instead of going down to the ocean in rivers.

The rich variety of valley-fill sediments reflects the wide range of environments that exist in most desert valleys. Streams that flow out of the mountains lay down deposits of sediments on the valley floor. Here and there, tracts of sand dunes slowly migrate across the valley floor, forming thick deposits of wind blown sand that other sediments will eventually bury. Most desert valleys have playa lakes in their lowest parts, shallow floods of salty and alkaline water where thick sequences of lake sediments accumulate. Deposits of peat laid down in marshy bogs around the edges of those playa lakes eventually turn into beds of coal.

Wind blown dust, pale gray silt blown off the deposits of desert sediment, covers much of the older basalt flows. In many areas it lies as a continuous blanket, in others as drifts huddled in lee places where the silt settled out of the wind. From an economic point of view, wind blown dust is the most important kind of desert sediment.

Desert soils naturally tend to be extremely fertile because heavy rains have never rinsed out their natural supply of soluble fertilizer nutrients. All they need to produce bountiful crops is irrigation water; it really does make the "desert bloom like a rose." Wind blown dust is especially nice because it may spread across lava flows in much less time than it would take them to weather into soil. The wind may even blow dust out of the desert into surrounding regions that naturally receive enough water to grow crops. Think of that as the reverse of irrigation: taking the soil to the water. That is what happened in the Palouse Hills.

The Silver City District

The old town of Silver City, now long abandoned to the desert, was once the center of a thriving mining district that produced large quantities of gold and silver. Prospectors found placer gold on Jordan Creek in 1863, inspiring the usual stampede of hundreds of unemployed men. Within a year, miners following the gold upstream found two large vein deposits on opposite sides of War Eagle Mountain, launching the district on a career of underground mining that continued long after Chinese miners washed the last yellow flakes out of the stream gravels.

Early underground mines worked fabulously rich deposits, enriched ores that contained metals weathered out of parts of the vein long since eroded away and then redeposited just below the surface. That kind of ore body contains the original metals content of the vein plus that of hundreds or thousands of feet of vein that formerly existed above the present ground surface. Those enriched ores make marvelously good mines, while they last.

Unfortunately, such rich secondary concentrations never continue to any great depth. Mines soon work through the bonanza into the unenriched primary vein below. When that happened at Silver City, the primary ore turned out to be not only much leaner than the enriched ore, but to have a much lower proportion of gold to silver. In 1875, that low grade ore combined with failure of the Bank of California to close the underground mines.

Discovery of several new veins launched a second period of optimistic mining in 1889. Production rapidly rose to levels comparable to those of the 1860s, continued high for about a decade, then went into a long decline that lasted until 1914. Desultory mining after that maintained a very low level of production that ended in early 1942. Many gold mines abruptly closed then after the industry was declared not essential to the war economy, thus making its miners subject to the draft.

Old mining districts die hard.

The Delamar Mine, which had produced about eight million dollars worth of silver and gold from underground workings between 1880 and 1914, went back into operation as an open pit in 1977. The mine is capable of producing more than two million ounces of silver per year, and has reserves sufficient to stay in production at that rate for many years, provided the price of silver is good. The ore is in Miocene rhyolite.

Idaho 51
Mountain Home—Nevada Line

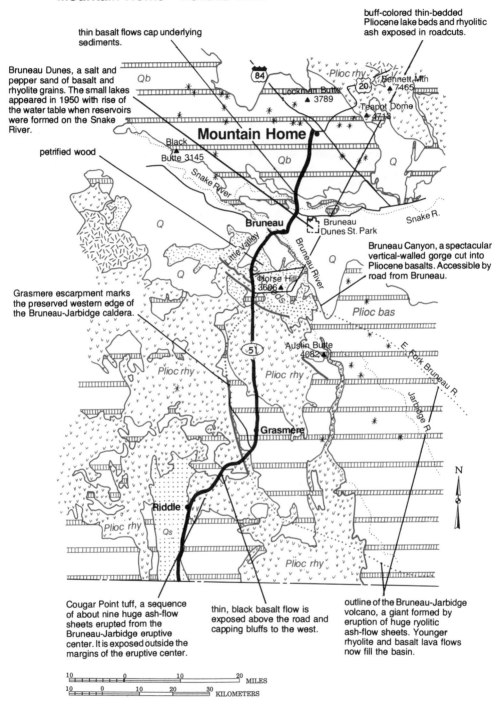

buff-colored thin-bedded Pliocene lake beds and rhyolitic ash exposed in roadcuts.

thin basalt flows cap underlying sediments.

Bruneau Dunes, a salt and pepper sand of basalt and rhyolite grains. The small lakes appeared in 1950 with rise of the water table when reservoirs were formed on the Snake River.

petrified wood

Qb

Plioc rhy

84

20

Lockman Butte
▲ 3789

Bennett Mtn
▲ 7465

Teapot Dome
▲ 4713

Mountain Home

Black
Butte 3145

Qb

Q

Snake River

Q

Bruneau

Bruneau
Dunes St. Park

Snake R.

Little Valley

Bruneau River

Horse Hill
3606 ▲

Qs

Q

Bruneau Canyon, a spectacular vertical-walled gorge cut into Pliocene basalts. Accessible by road from Bruneau.

Grasmere escarpment marks the preserved western edge of the Bruneau-Jarbidge caldera.

Plioc bas

51

Austin Butte
4082 ▲

E. Fork Bruneau R.

Plioc rhy

Plioc rhy

Jarbidge R.

Grasmere

N

Riddle

Plioc rhy

Qs

Plioc rhy

Cougar Point tuff, a sequence of about nine huge ash-flow sheets erupted from the Bruneau-Jarbidge eruptive center. It is exposed outside the margins of the eruptive center.

thin, black basalt flow is exposed above the road and capping bluffs to the west.

outline of the Bruneau-Jarbidge volcano, a giant formed by eruption of huge ryolitic ash-flow sheets. Younger rhyolite and basalt lava flows now fill the basin.

10   0   10   20 MILES
10 0 10 20 30 KILOMETERS

324

An eroded basalt volcano west of Bruneau, at the mouth of Sinker Creek. — U.S. Geological Survey photo by H.E. Malde

Idaho 51:
Mountain Home — Nevada Line
92 miles

Almost all the rocks exposed along this route are black basalt lava flows and much paler rhyolite ash flows of the Snake River Plain. Tell them apart by their colors. The only other rocks are deposits of light tan silt, much of it laid down in shallow lakes impounded behind lava flows. Except for some of the basalt near the opposite ends of the route, almost all the volcanic rocks erupted from the Bruneau-Jarbidge volcano.

The Bruneau-Jarbidge Volcano

Just west of Bruneau, the road crosses the Bruneau River, which most elegantly served the good cause of Idaho geology by slicing its deep canyon right through the old Bruneau-Jarbidge volcano. Rocks exposed in the canyon walls provide a marvelous cross-section through the volcano, one of the best sources of detailed information on the origin of the Snake River Plain.

Bruneau Canyon, a narrow saw slot of a gorge about 800 feet deep. — U.S. Geological Survey photo by H.E. Malde

Most of Highway 51 crosses the west side of the Bruneau-Jarbidge volcano. It is an enormous resurgent caldera that saw its period of most intense activity about 12 to 13 million years ago. Those eruptions seem to have been the start of the Snake River Plain, the first manifestation of the Yellowstone hotspot. The Bruneau-Jarbidge volcano probably began to form as more or less intact continental crust moved across the mantle site of the enormous crater that formed about 17 million years ago. It was at least as large as the modern Yellowstone volcano, probably considerably larger. It seems that the Yellowstone hotspot started full-sized, without the need for any warming up exercises.

The layers of rhyolite ash show that the Bruneau-Jarbidge volcano rapidly blew a series of at least 15 enormous ash flows, the Cougar Point tuffs, across much of southwestern Idaho and into nearby parts of Nevada and Oregon. Although the volume of all that rhyolite has not been estimated, it certainly amounts to at least dozens, probably hundreds, of cubic miles. The ground surface must have collapsed into a broad caldera basin as those enormous eruptions emptied the large magma chamber below. After the series of violent explosions had erupted the magma that carried a heavy charge of steam, dry rhyolite magma appeared quietly as thick lava flows and bulging lava domes.

Later basalt lava flows erupted from at least three dozen shield volcanoes more likely associated with Basin and Range faulting than with the hotspot. They filled any part of the caldera that the rhyolite lava flows may have left open. No hint of a basin survives in the modern landscape; no faint outline of the caldera shows through the opaque basalt flows.

Bruneau Dunes

Two grossly oversized sand dunes, the larger about 470 feet high, are the centerpiece of Bruneau Dunes State Park. Both stand in the Eagle Cove depression, an old meander scar of the Snake River. The Bonneville flood of some 15,000 or so years ago could hardly have failed to clean all sand out of the Eagle Cove depression, so these enormous piles of sand must have collected since then. The nearly incessant wind whipping across the Snake River Plain drops sand into the depression, grain by tedious grain. The lack of plants and the restlessly shifting sand surfaces with their moving ripples clearly indicate that the dunes are active.

The sand looks coldly gray from a distance. A close look through a magnifier shows a salt and pepper mixture of white quartz grains and black grains of basalt, in approximately equal proportions. The quartz grains came from weathering rhyolite.

The Bruneau dunes. —U.S. Geological Survey photo by H.E. Malde

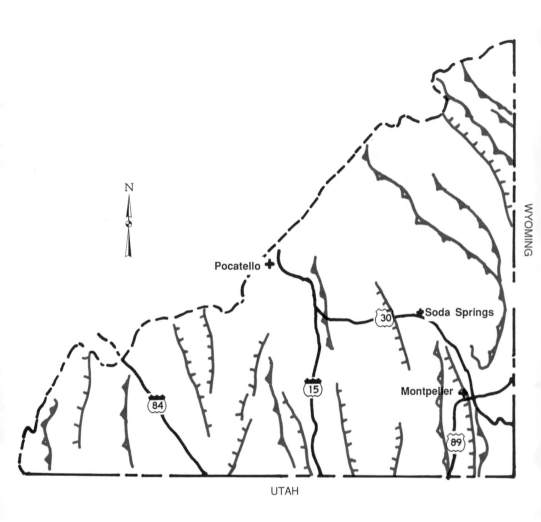

Roads covered in this section.

V

The Southeastern Mountains

Southeastern Idaho is a land of two generations of mountains: the Rocky Mountains with their overthrust belt, and the mountain blocks that the Basin and Range faults later chopped out of them. Although you now see them superimposed, the two sets of mountains differ radically in their ages, origins, and styles of deformation.

Before we plunge more deeply into those matters, we need to step back in time and look at some of the sedimentary rock formations that make up the mountains of southeastern Idaho. They accumulated over an enormous span of geologically quiet time before the more recent crustal movements rumpled, broke, and stacked them into mountains.

Sedimentary Rock Formations

Several of the mountain ranges in southeastern Idaho contain Precambrian sedimentary formations, of which too little is known. They could be younger or older than the Precambrian sedimentary rocks of central and northern Idaho, and they may record quite different environments and events. Knowing that those formations were laid down during the latter part of Precambrian time doesn't provide much help.

Precambrian time does, after all, embrace all the nearly four billion or so years of Earth's history that passed before animals began to flourish about 570 million years ago, at the beginning of Cambrian time. Even the latter part of Precambrian time, the interval known as Proterozoic time, embraces something like 1500 million years. That bottomless abyss of time contains plenty of room for the exquisitely slow accumulation of monstrously thick sequences of sedimentary rocks of quite different ages.

Geologists define the start of Paleozoic time with the abrupt appearance of numerous and varied Cambrian animals. No one knows why animal life appeared so suddenly then, and in such rich abundance and variety. Maybe that event had something to do with increasing oxygen content of the atmosphere, or decreasing carbon dioxide content, or perhaps both. Perhaps some other factor still undreamed of ignited that sudden abundance of life.

Several mountain ranges in southeastern Idaho contain thick sections of early Paleozoic sedimentary rocks, mostly limestones and sandstones, laid down during Cambrian and Ordovician time. Devonian time brought another period of widespread submergence, another substantial section of sedimentary rocks, including more limestone. Southern Idaho was submerged again during Mississippian time, when the Madison limestones formed. As Paleozoic time drew to a close during the Permian period, the Phosphoria formation was accumulating its vast and still mostly untapped store of phosphate rock in southeastern Idaho.

The most conspicuous of all the Paleozoic sedimentary rocks are the extremely thick Madison limestones. To judge from their fossils, those limestones must have accumulated in a shallow, tropical sea. Something like the Bahama Banks may be as close to an equivalent situation as the modern world offers. Limestone weathers very slowly in the dry climate of southern Idaho, so the pale gray Madison limestones make steep cliffs and high ridges wherever they come to the surface. They contribute a great deal to the scenery of southeastern and central, Idaho, very little to the economy.

The Phosphoria formation with its enormous reserves of phosphate rock may finally become the most valuable body of rock in Idaho. Its potential value certainly outweighs that of all the gold mines in the state, almost as certainly outweighs that of whatever unknown reserves of oil and gas may lurk undiscovered in the overthrust belt. The silver ores in the Coeur D'Alene district of northern Idaho provide the only visible competition for the honor.

Paleozoic time began with the abrupt appearance of abundant animal life on earth, and ended with the even more abrupt disappearance of most of the earth's animals, something like 90 percent of all the species then living. That horrendous cataclysm happened at the end of the Permian period, about 240 million years ago. To judge from the percentage of animal species that became extinct then, the terminal Permian catastrophe was even more drastic than the one that exterminated the Dinosaurs at the end of Cretaceous time, about 65 million years ago.

The overwhelming weight of evidence suggests that the Dinosaurs died in the aftermath of the explosive impact of a giant meteorite in western India. The Deccan flood basalts mark the spot. If a similar event ended Permian time, the enormous expanse of flood basalt lava flows in the Tungusska Plateau of western Siberia probably marks the spot. The Siberian basalts cover a considerably larger area than those in India, and they are about 240 million years old.

The years of the Dinosaurs began after the great extinction that ended the Permian period and ended about 65 million years ago with the catastrophic extinction of the dinosaurs and many other major groups of animals. That was Mesozoic time, which marched through the Triassic, Jurassic, and Cretaceous periods. During parts of all three of those periods, southern Idaho accumulated more layers of sedimentary formations; some were laid down in shallow sea water, others on land. We think those formations originally covered most of the region. Then the crustal movements that created the Rocky Mountains swept them all into the overthrust belt, and stacked them there.

The Rocky Mountains

The northern Rocky Mountains formed primarily through intrusion of large volumes of granite magma and massive eastward movement of the layered sedimentary rocks that had accumulated before all that activity began. The sedimentary rocks moved on faults that dip shallowly westward. In many cases, the movement left older formations lying on top of the younger ones — quite the reverse of the normal order. The only part of the older mountains that survives Basin and Range faulting in easily recognizable form is the overthrust belt.

The Idaho portion of the Rocky Mountain overthrust belt.

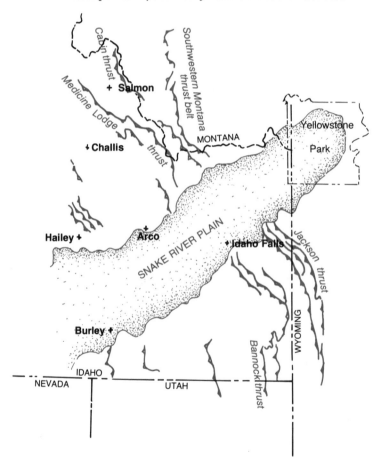

The northern Rocky Mountains specialize in overthrust faulting on a heroic scale. Great slabs of rock thousands of feet thick and dozens of miles across detached from the raised highland to the west and moved east, one after the other. Those thrust plates crumpled into folds as they moved, and piled onto each other as they came to rest in the eastern front of the mountain belt. The surfaces on which those slabs moved are faults, thrust faults if they dip fairly steeply, overthrust faults if they lie nearly flat.

This schematic drawing shows the general pattern of thrust faulting in the eastern part of the Rocky Mountains. Movement of older rocks onto younger ones is one of the main results.

Geologists differ radically and at times vehemently in their analysis of what moved those slabs of rock into the overthrust belt. Many stoutly maintain that a strong force applied from the west shoved the slabs of rock east. Others, including ourselves, argue that they simply slid east, moved downhill off the raised welt along the western edge of the continent under the pull of gravity. We find that many of the large overthrust faults in the northern Rocky Mountains contain granite that appears to have been molten magma when the fault moved. That probably solves the friction problem. A filling of molten magma would certainly lubricate a fault, making it easy for the rocks above it to glide downhill.

In a vaguely general way, the overthrust belt extends more or less continuously from the Canadian Arctic to Mexico. But the details of the rock formations involved and the way they moved and deformed vary greatly from one segment of the trend to another. The part of the overthrust belt in western

Wyoming and southeastern Idaho, for example, hardly resembles that in nearby southwestern Montana. But it does closely resemble the portion of the overthrust belt in southwestern Wyoming, which contains oil and gas.

Does Idaho Have Oil Wells in its Future?

Idaho is one of the few large states with no history of oil and gas production. None, but not for lack of bright hope, or of effort. People have been drilling here and there in Idaho for the past century.

In truth, most of Idaho has virtually no prospect of producing oil and gas, ever. Those commodities simply do not occur in Precambrian sedimentary rocks, basement rocks, metamorphic rocks, granite, and volcanic rocks. Idaho has more than its share of all those. But the state does have some thick sections of sedimentary rocks that were laid down in shallow sea water, and those may well contain oil and gas.

Idaho's main and most realistic hope for oil and gas production lies in its overthrust belt, where promising sedimentary formations are stacked on each other to great depth. They include the same formations that produce oil and gas in the southwestern Wyoming part of the overthrust belt. If good oil and gas fields exist in southwestern Wyoming, then why not in nearby southeastern Idaho? But finding them could be difficult. Consider what happened in Wyoming.

After Mormon emigrants found oil seeps in southwestern Wyoming in 1847, Brigham Young persuaded them to dig a well to produce oil to lubricate their wagon hubs and treat sores on their livestock. They dug, and the well did indeed produce some oil. Although most histories credit Colonel Drake with drilling the first producing oil well in western Pennsylvania in 1859, that hand dug hole in southwestern Wyoming may better deserve the honor.

The Mormon experience made it clear that southwestern Wyoming might be a good place to drill for oil. The first wildcat exploratory well was drilled in 1867. It was a dry hole, the first of almost 200 that would be drilled before the first really big discovery came, in 1977. The extremely complex folding and faulting in the overthrust belt makes it difficult to locate the right places to drill. Such a long series of dry holes is especially frustrating in an area where oil seeps leave no doubt that the stuff is there, somewhere. We think it likely that continued exploration will eventually develop oil and gas production from the Mesozoic formations of southeastern Idaho, but have no idea how long or how many dry holes that will take, or how much oil may exist.

Core Complexes

One inevitable effect of stretching the crust is to bring rocks that had been deep within it to the surface. In other words, separating the blocks of brittle rocks at the surface must create bare spots that expose the deep rocks beneath. Consider, for example, how the outer bark of many trees breaks into patches that separate as the tree grows beneath them, exposing deeper levels of bark beneath. Many geologists now suspect that those inevitable bare spots are the core complexes that exist here and there throughout the Basin and Range and Rocky Mountains. One of the best examples is in the Albion Range, south of Burley.

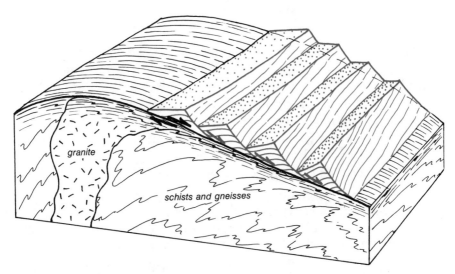

Idealized diagram of a typical core complex where stretching of the brittle rocks near the surface exposed deeper rocks.

The core of a typical core complex consists of a central area of rocks that formed at some considerable depth, generally granite or metamorphic rocks. A zone of very strongly sheared mylonite surrounds them, and the shallow rocks that formerly capped the deep rocks within are somewhere off to the side. The whole assemblage suggests that a piece of the upper crust was pulled off the deep rocks, letting them rise.

The question of when the many core complexes in the mountainous west formed remains a matter of continuing controversy. Almost certainly, most of them formed sometime after the main events that created the Rocky Mountains, sometime before the beginning of Basin and Range faulting. A date of around 20 to 40 million years ago seems to fit a great many. We are not sure of exactly what happened, or when.

The Basin and Range

After crustal movements created the Rocky Mountains about 70 to 90 million years ago, the slow processes of erosion began the infinitely tedious work of wearing them down. They were definitely showing signs of wear by sometime around 17 million years ago. Just then, the explosive impact of a giant meteorite in southeastern Oregon suddenly started a new and still continuing episode of mountain building that began to create the fault block mountains and valleys of the Basin and Range.

Basin and Range faulting started in southern Oregon and Nevada, and has migrated steadily eastward as the continent moves west. The eastern margin of the Basin and Range now lies along a line drawn along the Wasatch front in Utah, then north through southeastern Idaho and western Wyoming to Yellowstone Park, and beyond into Montana. All of southern Idaho west of that line lies within the Basin and Range, as does the Snake River Plain and the southern part of central Idaho.

Basin and Range faults trend generally north to northwest, dip steeply at the surface, and apparently flatten at depth. In every case, the block above the fault moves down, so each movement lengthens the crust. The curvature of the faults makes the blocks above them rotate as they move, raising one edge of the block and dropping the other.

336

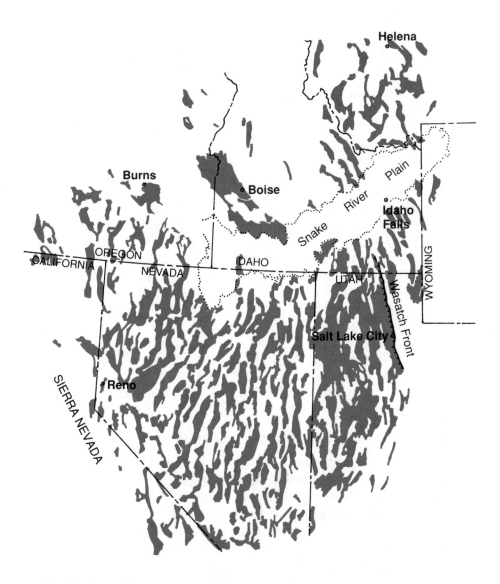

The Basin and Range region of the western United States extends from the Wasatch front in Utah west to the eastern margin of the Sierra Nevada in California

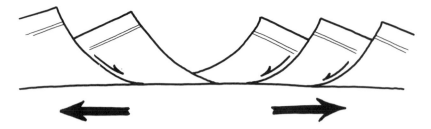

How movement on the curving Basin and Range faults lengthens the earth's crust and rotates the block of rock above the fault.

Each of the Basin and Range mountain blocks is a tilted slice of the Rocky Mountains pulled some distance from its original position. That is equally true of the valleys, but less obvious because the only rocks you can see in them are the much younger valley-fill sediments. Continuing fault movement pulls the mountain blocks farther away from each other, thus thinning the crust as the valleys grow wider. The Basin and Range has seen enough spreading to open a continental rift at least as wide as the Red Sea, if it were all concentrated along a single line. But the fairly rapid westward movement of North America over the rift in the mantle has distributed that spreading across the entire province.

The Ice Ages and Lake Bonneville

Few of the mountains in southeastern Idaho are high enough to make their own weather today; nor did they snatch enough moisture out of the clouds to maintain glaciers during the ice ages. Most of the region looks very much as it would had the ice ages never happened. Even so, the ice ages did make themselves felt.

One aspect of ice age climates was their very heavy precipitation, much more rain and snow than we now receive. All that water filled the Great Salt Lake Basin of northern Utah to form an enormous inland sea called Lake Bonneville. It flooded much of the northern half of Utah, some of eastern Nevada, and several of the broad Basin and Range valleys in southern Idaho. The water rose until it finally found an outlet through

338

overflow at the lowest point on the drainage divide. That was at Red Rock Pass, south of Pocatello. The overflow went down Marsh Creek to the Portneuf River, then down the Snake River.

Lake Bonneville began overflowing through Red Rock Pass on a formation of extremely hard and solid rock that maintained the lake at a constant level for a long time. The Bonneville shoreline precisely marks that elevation all around the Great Salt Lake Basin. A walk along that obvious horizontal bench on the mountain sides will take you past tracts of sand dunes, sandy beaches, deltas at the mouths of streams — absolutely everything you would expect along the shoreline of an inland sea, except the water.

The outlet stream through Red Rock Pass finally eroded through the hard rocks that had stabilized the lake at the Bonneville shoreline into a much softer formation beneath. Then the flow rapidly ripped some 300 feet of those soft rocks out of the pass, abruptly draining that much off the top of a lake that flooded an area a third the size of Utah. The modest stream that had been pouring through Red Rock Pass suddenly swelled to a torrent on a scale resembling that of the Amazon River. The Snake River canyon, much of which is about 600 feet deep, flowed more than brimful for several months as the lake level dropped. That was the Bonneville flood.

When Lake Bonneville finally stabilized at the level of the next formation of hard rock in Red Rock Pass, the Snake River Canyon had been thoroughly scoured all the way to its mouth. You can see scrubbed canyon walls, whirlpool eroded alcoves, and giant gravel bars hundreds of feet high all along the Snake River Canyon. And you can see the Provo shoreline that corresponds to the present level of Red Rock Pass all around the Great Salt Lake basin. It is precisely parallel to the Bonneville shoreline and some 300 feet below it.

As the climate became drier at the end of the last ice age, Lake Bonneville slowly dried up. Dozens of shorelines very faintly engraved on the slopes below the Provo shoreline record the slow stages of that long emptying. You can think of the Great Salt Lake as the last shrivelled remnant of Lake Bonneville.

Because it has no outlet stream to stabilize its level, the Great Salt Lake now rises during wet years and shrivels during droughts. A long period of much wetter climate than the region has known for the past 10,000 or so years would refill the basin back to the Provo shoreline. Then water would again spill through Red Rock Pass and down the valleys of Marsh Creek and the Portneuf River.

Except for the stretch near Malad Summit and that between Inkom and Pocatello, the highway follows broad structural valleys that dropped along Basin and Range faults. Most of the route crosses soft sediments and basalt that cover the older bedrock.

Mountains on both sides of the route are blocks raised along Basin and Range faults. They consist mostly of quite a variety of Paleozoic sedimentary formations, all much folded and broken along faults. Near Pocatello, the mountains also contain large volumes of older Precambrian sedimentary rocks. They were laid down sometime within the latter part of Precambrian time, but their exact age remains to be discovered.

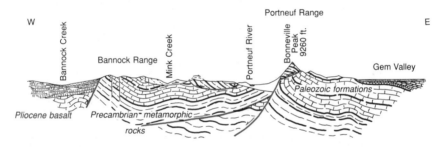

Section drawn across the line of Interstate 15 just south of Inkom. The faults that lie at low angles moved while the Rocky Mountains were forming. The steep faults belong to the Basin and Range.

The Missing Mesozoic Rocks

The considerable inventory of sedimentary formations in the mountains along Interstate 15 includes nothing deposited during Mesozoic time. That seems curious. Mountains of the overthrust belt just a few miles to the east contain enormous amounts of Mesozoic sedimentary rocks.

It is possible, of course, that the absence of Mesozoic formations could mean that no such rocks ever existed in this part of Idaho, but that hardly seems likely. Why should the area of their deposition end

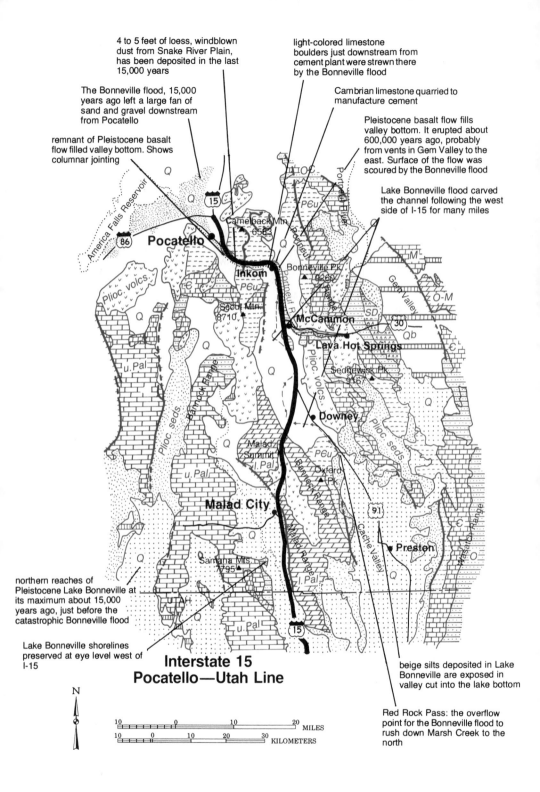

4 to 5 feet of loess, windblown dust from Snake River Plain, has been deposited in the last 15,000 years

light-colored limestone boulders just downstream from cement plant were strewn there by the Bonneville flood

The Bonneville flood, 15,000 years ago left a large fan of sand and gravel downstream from Pocatello

Cambrian limestone quarried to manufacture cement

remnant of Pleistocene basalt flow filled valley bottom. Shows columnar jointing

Pleistocene basalt flow fills valley bottom. It erupted about 600,000 years ago, probably from vents in Gem Valley to the east. Surface of the flow was scoured by the Bonneville flood

Lake Bonneville flood carved the channel following the west side of I-15 for many miles

Q

15

America Falls Reservoir

86

Pocatello

Camelback Mtn. 6583

Plioc. volcs.

Q

PЄu

Bonneville Pk. 9260

Gem Valley

O-M

Inkom

Scout Mtn. 8710

McCammon

30

SD

Qb

u. Pal

Bannock Range

Plioc. seds.

Lava Hot Springs

Sedgewick Pk. 9167

Q

Q

Plioc. volcs.

Downey

Plioc. seds.

Malad Summit

l. Pal

u. Pal

PЄu

Oxford Pk.

Bannock Range

91

Malad City

Q

Q

Cache Valley

Preston

Wasatch Range

Samaria Mts. 7795

l. Pal

northern reaches of Pleistocene Lake Bonneville at its maximum about 15,000 years ago, just before the catastrophic Bonneville flood

UTAH

u. Pal

15

Lake Bonneville shorelines preserved at eye level west of I-15

Interstate 15 Pocatello—Utah Line

N

beige silts deposited in Lake Bonneville are exposed in valley cut into the lake bottom

10   0   10   20 MILES
10   0   10   20   30 KILOMETERS

Red Rock Pass: the overflow point for the Bonneville flood to rush down Marsh Creek to the north so abruptly just a few miles to the east? It is also possible that Mesozoic formations did exist here until they were lost to erosion, but that also seems unlikely. Why should those rocks have been so completely eroded here, while they survived in the mountains farther east? We find it far easier to imagine that the Mesozoic formations were here until they moved east into the overthrust belt, uncovering the much older rocks you now see.

Basin and Range

All the rock formations in the mountains were tightly folded and much broken along faults while the Rocky Mountains formed some 70 or 80 million years ago. But the mountain ranges you see today did not form then. Erosion had reduced those older mountains more or less to nubbins, to a landscape of fairly low hills long before Basin and Range faulting began to transform the scene a few million years ago. Those faults are now slicing the old Rocky Mountains into the new fault blocks. The folds and faults within the mountains formed with the Rockies, but the broad outlines of the ranges and valleys belong to the Basin and Range.

Most of those younger faults trend generally north or northwest. Occasional earthquakes convince everybody that the faults are still active, that the mountains in this part of Idaho really are rising and the valleys dropping as the earth's crust slowly stretches. Near Malad Summit, the road crosses a large basalt lava flow that erupted during Pliocene time, doubtless in association with Basin and Range faulting. Watch for the black palisades of vertical columns.

The broad depression Interstate 15 follows between the Utah line and Malad Summit is a Basin and Range valley, as is another between the area near Downey and that near Inkom. The Portneuf River did not erode this part of its valley, but merely followed the dropping fault block. It did erode the much narrower stretch of its valley downstream from Inkom.

The Portneuf Canyon and Lava Flow

The road follows the broad floor of the Portneuf Valley all the way between its junction with Highway 30 and Pocatello, on the Portneuf lava flow. The basalt is much better exposed than it once was, now that the Bonneville flood has efficiently scoured deep deposits of wind blown silt off its surface. The flood also ripped through the flow along much of that route to create the valley floor in which the Portneuf River now flows. Remnants of the flow remain as broad benches along both sides of the valley. Watch for the distinctive rows of vertical basalt columns along the edges of those benches.

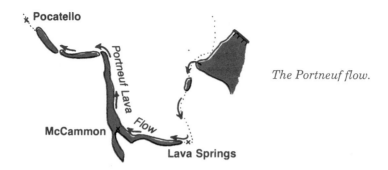

The Portneuf flow.

Age dates show that the Portneuf basalt flow erupted about 600,000 years ago. Although the connection is not continuously exposed, it seems very likely that it came from the Gem Valley lava field, about 17 miles east of McCammon. If so, the farthest end of the flow near Pocatello is at least 30 miles from its source. Imagine that broad river of glowing lava pouring down the valley floor beneath clouds of smoke rising from trees burning in its path.

The Portneuf River cuts right across the grain of both the topography and the bedrock geologic structure where it slices through the Portneuf Range between Lava Hot Springs and McCammon, again in its canyon through the Bannock Range between Inkom and Pocatello. Evidently, the river is coping with the moving fault blocks by eroding its bed fast enough to keep pace with their movements, maintaining its course through the hills as they rise across its path.

A big lava flow fills the flat floor of Portneuf Canyon. The great pioneer photographer W.H. Jackson exposed a wet plate negative more than a century ago to take this picture for the U.S. Geological Survey.

344

Red Rock Pass, the deeply eroded spillway that drained Lake Bonneville from the Bonneville to the Provo shoreline. The floor of the pass is exactly at the level of the Provo shoreline.

The Great Flood

Along much of the way south of Malad Summit, the highway follows a route that was under water when Lake Bonneville existed. Unconsolidated tan sediments exposed here and there beside the road were deposited on the floor of the big lake. Although the road does cross Lake Bonneville shorelines between Malad Summit and the Utah line, they are extremely difficult to spot. This was a sheltered bay where wave action was too weak to cut a prominent shoreline. People travelling south into Utah can watch for clearly visible shorelines on the face of the Wasatch Range east of the highway, especially between Brigham City and Provo.

Malad Summit is well above the higher Bonneville shoreline, and shows no evidence that water ever flowed through it, no sign of erosion. Red Rock Pass, which Highway 91 crosses about nine miles east of Malad Summit, appears to have been the only place where Lake Bonneville overflowed. Watch when you go through Red Rock Pass for the broad hay meadows along several miles of its floor, just below the level of the road. Those are in the floor of the valley that overflow from Lake Bonneville eroded through the pass, at the level of the Provo shoreline.

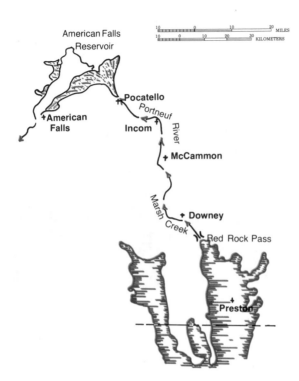

The line of the highway and that of the flood scoured valleys of Marsh Creek and the Portneuf River.

When Lake Bonneville drained through Red Rock Pass, the water flowed down Marsh Creek into the lower Portneuf River, then on into the Snake River. For a long time, hard rock in the floor of the pass resisted erosion, stabilizing the lake at the level of the Bonneville shoreline. During those years, the overflow simply maintained a normal stream that flowed north through Red Rock Pass. After erosion finally cut through the hard rock, into a much softer formation beneath, rapid erosion of the softer rocks caused the catastrophic Bonneville flood.

While the lake was rapidly draining from the Bonneville to the Provo shoreline, the discharge through the valleys of Marsh Creek and the Portneuf River was approximately 15 million cubic feet per second, about one-third of a cubic mile per hour. Watch for the scoured and scrubbed look of the landscape where the highway crosses the path of the flood between Downey and McCammon, and for the

346

View from Red Rock Pass down the valley of Marsh Creek.
—U.S. Geological Survey photo by R.W. Stone.

scrubbed basalt here and there along the Portneuf River between McCammon and Pocatello. Watch also for giant ripples, enormous waves of sand and gravel shaped just like the little sand ripples you see in stream beds, but ten feet or more high and tens of feet from crest to crest.

After the Bonneville flood, the lake level again stabilized, this time at the elevation of the modern floor of Red Rock Pass. Wave action along the new shoreline cut the Provo shoreline, which is not nearly so strongly developed as the higher Bonneville shoreline. Evidently, the lake stood at its lower elevation for a relatively short time before its level began to drop through evaporation in the drier climate that followed the end of the last ice age.

Perlite

One of Idaho's few large deposits of good quality perlite exists a few miles west of the highway about 25 miles north of Malad City. Perlite is a variety of rhyolite glass that contains a small amount of water. When the rock is broken into small pieces and then heated, the water turns to steam and the pieces pop like kernels of popcorn. Expanded perlite has excellent insulating and soundproofing qualities. It is widely used in making lightweight concrete blocks, ceiling tiles, and wall board. Gardeners use the little white pellets as a soil conditioner. Look for them in bags of potting soil.

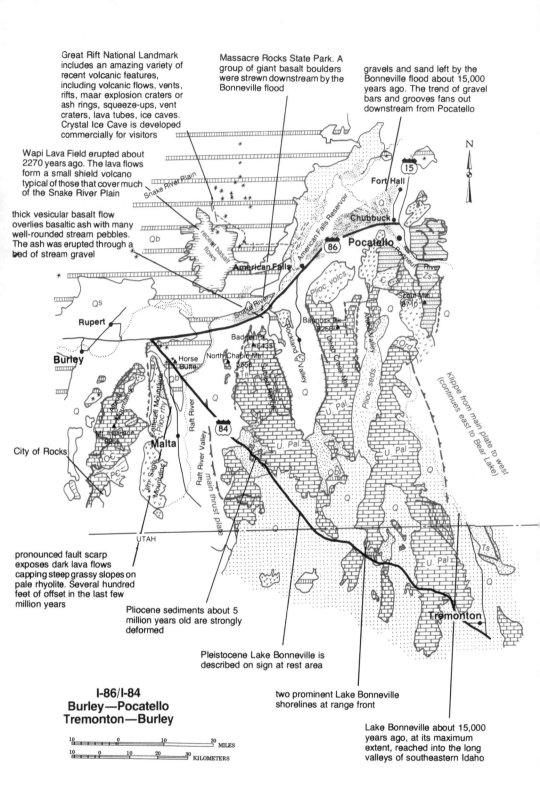

Great Rift National Landmark includes an amazing variety of recent volcanic features, including volcanic flows, vents, rifts, maar explosion craters or ash rings, squeeze-ups, vent craters, lava tubes, ice caves. Crystal Ice Cave is developed commercially for visitors

Wapi Lava Field erupted about 2270 years ago. The lava flows form a small shield volcano typical of those that cover much of the Snake River Plain

thick vesicular basalt flow overlies basaltic ash with many well-rounded stream pebbles. The ash was erupted through a bed of stream gravel

Massacre Rocks State Park. A group of giant basalt boulders were strewn downstream by the Bonneville flood

gravels and sand left by the Bonneville flood about 15,000 years ago. The trend of gravel bars and grooves fans out downstream from Pocatello

N

Snake River Plain

Fort Hall

15

Chubbuck

Qb

86 Pocatello

American Falls

recent basalt flows

Qb

Plioc. volcs.

Pocatello

River

Zs

Qs

Rupert

Shake River

Rockland Valley

Bannock Pk. 9256

Scott Mtn. 8710

Q

Badger Pk. 6435

North Chapin Mtn 5856

Deep Creek Mts

Arbon Valley

Q

Qrd

Burley

Horse Butte

Qb

Sublett Range

Plioc. rhy.

Plioc. seds.

Klippe from main plate to west (continues east to Bear Lake)

City of Rocks

Albion Mountains

Cotterell Mountains

Jim Sage Mountains

Raft River

Raft River Valley

84

Q

U. Pal.

U. Pal.

U. Pal.

Q

Malta

Mt. Harrison 9265

Q

main thrust plate

Q

UTAH

pronounced fault scarp exposes dark lava flows capping steep grassy slopes on pale rhyolite. Several hundred feet of offset in the last few million years

Pliocene sediments about 5 million years old are strongly deformed

U. Pal.

Q

Ts

Tremonton

Pleistocene Lake Bonneville is described on sign at rest area

two prominent Lake Bonneville shorelines at range front

**I-86/I-84
Burley—Pocatello
Tremonton—Burley**

10   0   10   20 MILES
10   0   10   20   30 KILOMETERS

Lake Bonneville about 15,000 years ago, at its maximum extent, reached into the long valleys of southeastern Idaho

Interstate 84:
Interstate 86 — Utah Line
48 miles

The intersection with Interstate 86 is near the southern edge of the Snake River Plain, where basalt lava flows form the bedrock. The rest of the route passes through Basin and Range country, a land of isolated mountain ranges rising above broad basins. Both ranges and basins are fault blocks chopped out of the geologically intricate Rocky Mountains. Our cross-section shows big thrust plates lying on nearly horizontal faults; those are Rocky Mountain structures. The steeply dipping faults such as those that bound the Raft River Valley belong to the Basin and Range; they cut the older faults.

Like most roads in the Basin and Range, this highway tends to follow the broad valleys and to avoid the mountains. About all you can see at the surface in the basins is fairly recent sedimentary deposits. You gaze across the broad valley floors at the varied and complexly deformed bedrock formations in the distant mountains. About half the northern part of the route passes through the Raft River Valley with the Albion Mountains on the west, the Sublett Range on the east. A low pass in the Sublett Range, probably an incipient fault block basin, separates the Raft River Valley from another large valley near the Utah line.

Volcanic Rocks

As a rule of thumb, geologists broadly classify the younger lava flows on the Snake River Plain into loose and approximate age groups according to the proportion of rock still without soil cover. The youngest flows are more than 75 per cent exposed, those of intermediate age between 25 and 75 per cent bare rock, and the oldest are less than 25 per cent exposed. Still older lava flows are completely covered with soil and plants.

The remarkably straight Horse Butte fault scarp breaks the Snake River Plain near the north end of the route.

The northern end of the route crosses large basalt flows. They probably erupted from the small volcano with a beacon on its summit that rises about a mile east of the highway, approximately three miles south of Interstate 84. Less than one fourth of the surface of these flows is bare rock, so they belong to the oldest group of younger flows, which probably means they are some hundreds of thousands of years old. Their age suggests that these flows must be associated with Basin and Range faulting rather than with the Yellowstone hotspot, which passed this way millions of years before they erupted.

A prominent fault scarp west of the highway near Horse Butte exposes a thin cap of black basalt lava flows lying on pale rhyolite ash, the typical sequence in the Snake River Plain. The fault scarp is so sharp and shows so little erosional dissection that it can't be very old. Like most Basin and Range faults, this one trends approximately north.

Raft River Valley

The Raft River Valley lies between the Jim Sage and Cottrell Mountains on the west, and the Sublett and Black Pine Ranges on the east. Its southern boundary is the Raft River Mountains of northern Utah, one of the few ranges in the Basin and Range that trend from

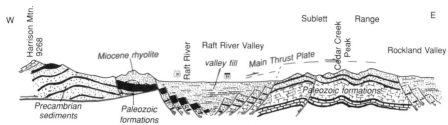

Section across the Raft River Valley about 18 miles south of junction of Interstate 86.

east to west. Perhaps it is a remnant of the high flank of the Snake River Plain left by passage of the Yellowstone hotspot.

Like all the fault block basins in this region, the Raft River Valley contains a deep fill of sedimentary and volcanic rocks, which appear to have accumulated during latest Miocene and Pliocene time. They include deposits of silt, sand, gravel, mudflows, lake bed sediments and volcanic ash, along with some locally erupted rhyolite and basalt. In the modern world, deep deposits of sediments typically accumulate in desert valleys so dry that no river drains them. So it seems most unlikely that the Raft River could have existed while the valley fill sediments were accumulating. They almost certainly record a Pliocene climate much drier than the one that now prevails.

All the volcanic activity evidently heated the rocks beneath the valley floor. Relatively shallow wells drilled in the Raft River Valley have produced amazing volumes of very hot water and steam. They leave no doubt that the valley-fill sediments hold substantial geothermal energy reserves. If the Raft River Valley were not in such a thinly populated region, all that heat would by now be generating electricity. Wait a few years.

The Sublett Range

Rocks in the Sublett Range include a wide variety of sedimentary formations. Most were laid down during the latter part of Paleozoic time, between 300 and 200 million years ago, in round numbers. The only area where the road actually crosses those older rocks is near the pass south of the Raft River Valley.

All those sedimentary rocks were tightly folded and broken along faults while the Rocky Mountains were forming, 70 to 90 million years ago. Large slabs of rock that moved east along the nearly horizontal faults probably came from at least as far west as the Albion Range. All those crustal movements probably left this a mountainous region. At least 60 million years of erosion then reduced those contorted rocks to a landscape of modest but rugged hills.

Much younger volcanic rocks covered the eroded older rocks in parts of the Sublett Range as Basin and Range faults began to chop the deformed rocks into the crustal blocks that are the modern mountain ranges. In this area, that most recent stage of mountain building began about eight million or so years ago, and still continues. The Basin and Range faults are certainly moving and more basalt eruptions should surprise no one.

Of Fans and Climates

On both sides of the pass through the Sublett Range, the road crosses big alluvial fans, now considerably dissected by streams. Alluvial fans typically form in very dry regions, deserts, where the sudden floods of runoff water that follow occasional heavy rains carry enormous loads of sediment. Those brief floods leave much of that sediment behind in big alluvial fans, which are shaped like segments of cones laid on the flanks of the mountains. Each flood of sediment fills any low places on the fan surface to build the smoothly conical shape.

Wetter climates maintain enough plant cover to prevent the catastrophic soil erosion and enormous floods of surface runoff typical of deserts. Leaves intercept falling rain drops so they can't splash the soil when they hit, thus greatly reducing the rate of erosion. Meanwhile, burrowing animals that live among the plants open pore spaces that permit water to soak into the ground, which reduces the amount of surface runoff. That also reduces the rate of soil erosion.

By reducing the rate of soil erosion, plants tend to foster clear streams that are more likely to erode their beds than to deposit sediment in them. As they increase the proportion of rain that soaks into the ground, plants expand the reservoir of stored ground water available to keep the streams flowing through dry seasons. The result is perennial streams that erode valleys in fans, instead of covering them with new blankets of sediment.

So the big alluvial fans now much dissected by streams are evidence that the amount of rainfall has varied greatly during the fairly recent geologic past. It seems likely that the fans grew during dry periods, which were probably interglacial periods between ice ages. Ice ages probably brought the wet periods in which streams dissected the fans.

The Albion Range,
a Metamorphic Core Complex

Like the Sublett Range, the Albion Range consists mostly of Paleozoic sedimentary formations that were tightly deformed during formation of the Rocky Mountains. Then, about eight or ten million years ago, the Yellowstone hotspot blanketed them with volcanic rocks as it passed to the north. Now, Basin and Range faults are chopping those old mountains with their blankets of rhyolite into blocks, and pulling the blocks apart to make broad valleys separated by mountain ranges.

Some of the sedimentary rocks in the Albion Range are much older than any in the Sublett Range. The Albion Range contains Cambrian formations deposited in earliest Paleozoic time, nearly 600 million years ago, as well as a series of Precambrian sedimentary formations. Neither the exact age of those more ancient rocks nor their relationship to the Precambrian sedimentary formations of central and northern Idaho is known.

The Albion Range is best known among geologists for its several areas of profoundly sheared mylonites. They formed along fault zones at a level so deep within the continental crust that the rocks were hot enough to deform by flowing like modelling clay, instead of through breakage. Those mylonites appear to outline an area of deep seated rocks that rose as the rocks that formerly covered them moved off along the mylonite fault zones, as though they were blankets drawn off a bed.

The Sublett and Black Pine Mountains along the east side of the Raft River Valley probably contain the rocks that moved off the mylonite. If so, those rocks moved east at least 50 miles. Geologists call areas of deep seated rocks exposed through unloading of their cover metamorphic core complexes. The age of these is uncertain, probably somewhere between 20 and 40 million years. Mylonites and metamorphic core complexes are hard to date.

Silent City of Rocks

The southern part of the Albion Range contains a large mass of granite emplaced during Tertiary time, long after the Rocky Mountains formed. That granite erodes into a fabulous landscape, a fairyland of sculptured rocks called the Silent City of Rocks.

Silent City of Rocks was once one of the busiest crossroads in Idaho, a famous landmark of the Oregon Trail. This is where the Hudspeth Cutoff to California branched off the main trail. It is now a remote

Weathered pinnacles of granite in the City of Rocks. Many pioneers travelling the Oregon Trail left their names carved in these weirdly sculptured outcrops.

outpost at the end of six miles of unpaved side roads from the nearest community, Almo. Here, outcrops of the Cassia granite are eroded, sculptured might be a better word, into a bizarre array of pinnacles, rounded domes and boulders scattered across a basin that covers about nine square miles. It is a fairyland of fantastically weathered rock. Granites rather commonly erode into such forms, but rarely as completely or as intricately as in the Silent City of Rocks. Here you see the essence of weathered and eroded granite in its ultimate expression.

It seems clear that erosion of such granite domes and boulders happens in two stages. First, the granite weathers into soil along deep fractures while the rock between the fractures remains perfectly fresh. Then, something, probably fire or drought, destroys the plant cover and the exposed soil erodes, leaving the unweathered granite between fractures standing above the ground surface. So the sculptures we see formed as the granite weathered beneath the surface; then appeared as that surface eroded.

The sculptured forms are invariably rounded, never angular. That reflects the tendency of rocks to weather where water seeps into them along fractures. Edges and corners where fractures intersect weather most rapidly because seeping water attacks them from two or three directions. Meanwhile, rounded masses of rock remain fresh and unweathered in the centers of the original square blocks. Those corestones are the round boulders we see at the surface after erosion removes the soil that formed along the fractures. They were rounded in place as they weathered, not by stream transport.

Continued weathering since the rocks were exposed has added intricate details, which depend heavily upon a brownish crust of iron oxides called desert varnish, because it forms most conspicuously in arid regions. The extremely insoluble iron oxide varnish protects the rock within from weathering. Where something breaks the varnish, the rock beneath rapidly weathers into hollows that give some of the pinnacles the look of fairyland castles complete with windows and doors.

The Cassia granite was emplaced as molten magma long after the Rocky Mountains formed, during early Tertiary time. Like many Idaho granites of that age, it contains large pegmatite dikes in which the rock is so coarsely crystalline that individual mineral grains exist as single crystals that average several inches across — a few pegmatites are even coarser. Several small mines in the area about a half mile southwest of the center of the Silent City of Rocks produced modest amounts of feldspar from pegmatites. Most of it was used in making porcelain.

View looking south from Interstate 84 in Utah, eight miles south of the Idaho border. The horizontal line about halfway up the mountain front is the Bonneville shoreline.

Lake Bonneville

Lake Bonneville flooded into the southern part of the route, into the broad valley just north of the Utah line. Watch for shorelines on the mountains: the Bonneville shoreline at an elevation of about 5100 feet, the Provo shoreline about 300 feet lower. Neither is easy to see from the Idaho portion of this road. Both become more easily visible farther south, in Utah. Watch also for the tan lake silts in the flat floor of the valley near the Utah line. The Raft River Valley farther north is low enough that it would have flooded, had the intervening passes been a bit lower.

The former extent of Lake Bonneville in southern Idaho.

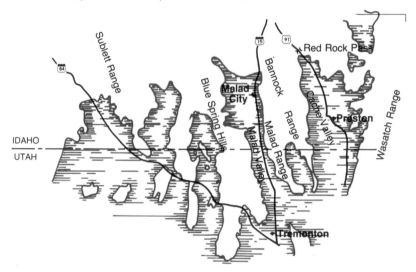

View along the Snake River east of Idaho Falls. Basalt fills the valley floor.

U.S. 26:
Idaho Falls — Wyoming
64 miles

Idaho Falls is near the eastern edge of the Snake River Plain, where the basalt flows lap onto the Caribou Mountains. Highway 26 crosses a short stretch of basalt, quite a large area of rhyolite between Idaho Falls and Swan Valley. Both belong to the Snake River Plain.

Between Swan Valley and the Wyoming line, the highway follows the Swan and Grand valleys between the Caribou Range on the southwest and the Big Hole and Snake River mountains to the northeast. The valleys are in a dropped Basin and Range fault block chopped into the overthrust belt.

The Overthrust Belt

The eastern edge of the Basin and Range is just beginning to dismember the Idaho portion of the overthrust belt into moving fault blocks. Rocks there are a thick stack of Paleozoic and Mesozoic sedimentary formations that moved northeast along thrust faults

357

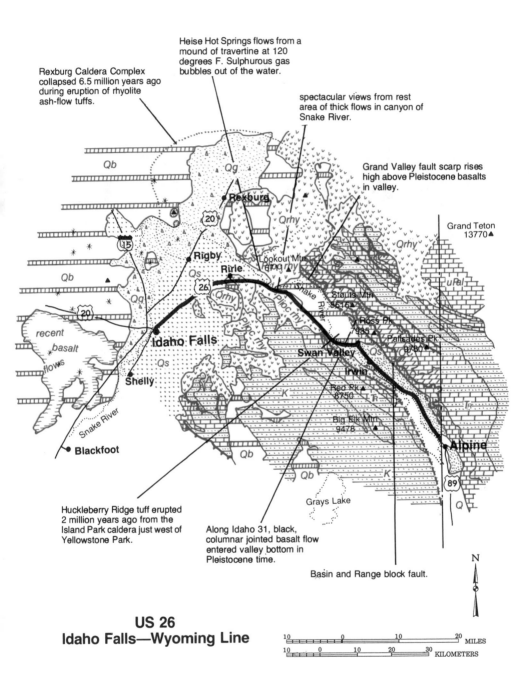

Heise Hot Springs flows from a mound of travertine at 120 degrees F. Sulphurous gas bubbles out of the water.

Rexburg Caldera Complex collapsed 6.5 million years ago during eruption of rhyolite ash-flow tuffs.

spectacular views from rest area of thick flows in canyon of Snake River.

Grand Valley fault scarp rises high above Pleistocene basalts in valley.

Grand Teton 13770▲

recent basalt flows

Huckleberry Ridge tuff erupted 2 million years ago from the Island Park caldera just west of Yellowstone Park.

Along Idaho 31, black, columnar jointed basalt flow entered valley bottom in Pleistocene time.

Basin and Range block fault.

N

US 26
Idaho Falls—Wyoming Line

Section across the Caribou and Snake River ranges about five miles north of Swan Valley. The thrust faults that carried the tightly folded sedimentary formations east all dip down to the southwest. The much younger faults that outline the Swan Valley belong to the Basin and Range.

some unknown distance, at least tens of miles as the Rocky Mountains formed during late Cretaceous time. Their long journey into the overthrust belt left the rocks much battered, broken along faults and bent into folds.

The basic structure of the overthrust belt is a series of several great slices of sedimentary rock, thrust plates, each many thousands of feet thick, that moved generally northeast on faults that carried them up and over the rocks in their path. In many places, older rocks now lie on top of younger formations in absolute reversal of the prescribed common sense order: The oldest sedimentary layers are supposed to lie at the bottom of the stack. Rocks within each thrust plate are tightly folded, especially near the big faults.

Streams have since cut their valleys through the thrust plates, dissecting them into eroded mountains in which the valleys tend to follow the sections of less resistant rock. The overthrust belt faults and folds all trend northwest, so that is the general trend of the eroded stream valleys. The Basin and Range faults that are now beginning to cut the overthrust belt trend in the same direction. So valleys trend northwest in the Idaho overthrust belt regardless whether they were eroded into less resistant rock formations or dropped along Basin and Range faults. That persistent trend of the valleys, and of the ridges that separate them, gives the region its strong topographic grain.

Between the area just east of Rigby and the Wyoming line, the highway follows the Snake River through the Swan and Grand valleys, a large Basin and Range fault block. Basalt lava flows and rhyolite probably associated with the Basin and Range flooded the nearly flat floor of the Swan Valley. Valley-fill sediments floor the Grand Valley.

Rocks in the ridges of the Caribou Range southwest of the Grand and Swan valleys are Mesozoic formations deposited during Triassic and Jurassic time. Those in the Snake River Range northwest of the highway are much older formations deposited during Paleozoic time. The Big Hole Range contains both Mesozoic and Paleozoic formations.

Gold

Early prospectors discovered placer gold deposits in the streams that drain from Caribou Mountain, near the Wyoming line. The first discoveries came in 1870, and most of the gravel was washed during the 1880s and 1890s. No actual records are available, but some estimates suggest that the total production may have been a few million dollars. That seems a paltry return, considering the large investment of hard labor the old placer workings so obviously record.

The gold came from Caribou Mountain. It contains a swarm of igneous intrusions composed of a variety of uncommon rock types, all of which are peculiar in containing considerably more than the normal amount of potassium. Such magmas commonly import gold into the rocks they intrude, so it is no surprise to find old hard rock gold mines and prospects on Caribou Mountain. We haven't seen any estimates of total production from the underground workings, but none of the mines have the look of a property that produced much.

U.S. 30:
Interstate 15 — Wyoming
84 miles

From a geologic point of view, this is one of the most varied and interesting drives in Idaho, also one of the more complex. The broad basins are fairly simple, at least on the surface. They contain unconsolidated sedimentary formations that accumulated during Pliocene and Pleistocene time, as they dropped along faults. The rocks in the mountains vary greatly from one range to the next. Let's try to sort things out by beginning with a brief tour of the rocks along each part of the drive.

A Geologic Tour

In the western part of the route, between Interstate 15 and Bancroft, the road passes through mountains composed of folded sedimentary rocks so intricately shattered along faults that detailed geologic maps look like a hammered piece of glass. Most of those rocks are Paleozoic sedimentary formations deposited between 600 and 200 million years ago. The same mountains also contain some much older sedimentary formations deposited during Precambrian time, but none of those lie near this highway.

Between Bancroft and the intersection with Highway 34 at Soda Springs, the road follows a perfectly straight line across the nearly level expanse of the Gem Valley, a dropped Basin and Range fault block. Bedrock in the valley floor is basalt lava flows, well covered with soil and plants. The basalt magma almost certainly formed as Basin and Range faulting reduced pressure on the hot rocks in the upper mantle by stretching the earth's crust.

Hills east of the Gem Valley consist mostly of early Paleozoic sedimentary formations so tightly deformed that you need a magnifying glass to study the more detailed geologic maps. Most of those formations are limestones deposited in shallow sea water during Cambrian and Ordovician time, between 500 and 600 million years ago, in round numbers. Similar rocks form most of the hills north and south of the road between Soda Springs Reservoir and Soda Springs.

The highway builders followed the Bear River through a long series of valleys between Soda Springs Reservoir and Montpelier. The generally straight northwest trend of these valleys follows the trace of the Bannock thrust fault, a Rocky Mountain structure. The valley

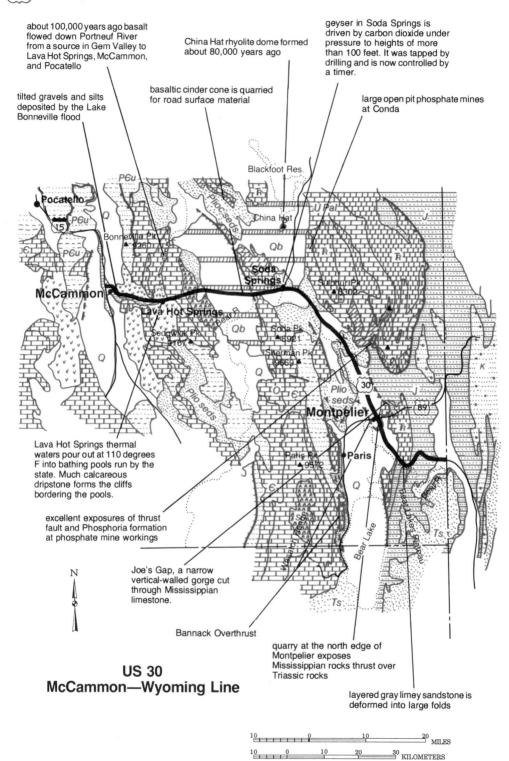

about 100,000 years ago basalt flowed down Portneuf River from a source in Gem Valley to Lava Hot Springs, McCammon, and Pocatello

China Hat rhyolite dome formed about 80,000 years ago

geyser in Soda Springs is driven by carbon dioxide under pressure to heights of more than 100 feet. It was tapped by drilling and is now controlled by a timer.

basaltic cinder cone is quarried for road surface material

tilted gravels and silts deposited by the Lake Bonneville flood

large open pit phosphate mines at Conda

Blackfoot Res.

Pocatello

PЄu

PЄu

PЄu

Q

JЄ

PЄu

Plio seds

China Hat

U Pal

J

Bonneville Pk
9260

Qb

McCammon

Soda
Springs

Lava Hot Springs

Sulphur Pk
8302

Sedgwick Pk
9187

Qb

Soda Pk
8921

Q

Plio seds

Sherman Pk
9669

Q

Ѐ

Plio
seds

30

Montpelier

89

K

J

Ѐ

Lava Hot Springs thermal waters pour out at 110 degrees F into bathing pools run by the state. Much calcareous dripstone forms the cliffs bordering the pools.

Paris Pk
9572

Paris

Q

Wasatch Range

excellent exposures of thrust fault and Phosphoria formation at phosphate mine workings

Bear Lake

Bear Lakes Plateau

Ts

N

Joe's Gap, a narrow vertical-walled gorge cut through Mississippian limestone.

Ts

Bannack Overthrust

quarry at the north edge of Montpelier exposes Mississippian rocks thrust over Triassic rocks

US 30
McCammon—Wyoming Line

layered gray limey sandstone is deformed into large folds

10   0   10   20 MILES

10   0   10   20   30 KILOMETERS

route puts most of the contorted bedrock in the mountains on either side out of sight, but you can see some of the rocks that fill the valley floor.

Most of the route past Soda Springs Reservoir and Soda Springs crosses basalt lava flows, probably an extension of the Gem Valley lava field. About five miles east of Soda Springs, the road crosses a patch of rhyolite, also part of the same volcanic field. Along the rest of the route between Soda Springs and Montpelier, the road crosses a variety of fairly young sedimentary deposits, mostly desert alluvial fans, probably no more than a couple of million years old.

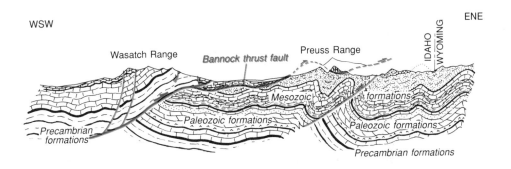

Section across the line of the highway about midway between Soda Springs and Montpelier. Older rocks west of the highway rode over the younger ones to the east along a big overthrust fault.

Bedrock in the mountains on either side of the road between Soda Springs and Montpelier is a complicated mess, part of the overthrust belt. Mountains in the Preuss Range east of the road consist mostly of tightly folded sedimentary formations deposited during late Paleozoic and early Mesozoic time, between about 280 and 120 million years ago. And there are small areas of pale gray Madison limestone deposited about 340 million years ago. Bedrock in the Wasatch Range west of the road consists mostly of much older sedimentary formations deposited during early Paleozoic time, between about 450 and 600 million years ago. Those rocks rode east on the Bannock thrust fault that carried them over the younger formations.

The short drive between Montpelier and the Wyoming line crosses more of the overthrust belt. Rocks in this area are Mesozoic sedimentary formations that crumpled into tight folds as they moved east along big thrust faults.

Unloading the Rocks West of the Overthrust Belt

Mesozoic sedimentary formations are just as conspicuously absent from the mountains west of the overthrust belt as they are conspicuously present in the overthrust belt. If it were possible to pull those formations out of the overthrust belt, shake out their wrinkles, and lay them flat, they would stretch halfway across southern Idaho. We think that is exactly where the Mesozoic sedimentary formations were until about 70 million years ago, when they moved east along thrust faults into the overthrust belt.

If thousands of feet of Mesozoic sedimentary rocks moved off the mountains west of the overthrust belt, then the earth's crust there must have risen as it shed that heavy burden. The crust, after all, floats on the deeper rocks beneath in about the same way that an air mattress floats on a lake. Both sink if loaded; rise if unloaded. We suspect that much of the intricate faulting in the ranges west of the overthrust belt may have formed as that area rose after losing its cover of younger rocks.

Basin and Range Faulting

This part of Idaho is near the eastern edge of the Basin and Range, where the continental crust began riding over the spreading ridge in the mantle within the last few million years. The fault block valleys and mountains are just beginning to form. Most of the Basin and Range volcanic activity is very recent, and still continuing.

The fault block ranges of southeastern Idaho continue to move a bit farther apart, and the fault block valleys between them grow a bit wider. Every year adds more acreage to southern Idaho, a little more distance between towns that lie east or west of each other. That continuing extension makes it more easily possible to imagine that the Mesozoic formations in the overthrust belt may have come from considerably farther west than the mountains near Pocatello. They didn't have nearly as far to come 70 million years ago as they now would.

Lava Fields

The Blackfoot lava field lies north of Soda Springs, around the Blackfoot Reservoir; the Gem Valley lava field a few miles west of town. Both volcanic centers consist dominantly of basalt cinder cones and lava flows. Several rhyolite domes rise as isolated hills above the nearly level sea of basalt.

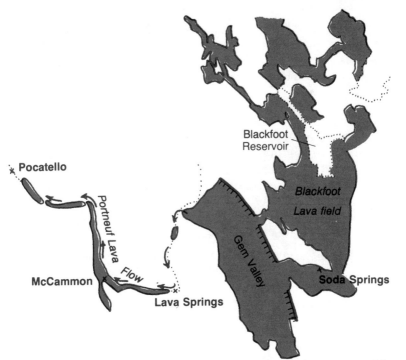

The Blackfoot lava field, the Gem Valley lava fields, and the Portneuf flow.

The thick blanket of soil and wind blown dust that covers much of the Blackfoot and Gem Valley lava fields clearly shows that most of the eruptions happened a long time ago. Age dates range from 50,000 to 1.4 million years, but some of those figures must be wrong. A million and a half years seems an unreasonably long period for activity to continue in a small lava field. Sometime around 600,000 years ago is probably a good estimate for the age of most of the eruptions.

The eroded edge of the Portneuf flow exposed beside U.S. 30 near McCammon.

365

Many of the cinder cones and rhyolite domes align in neat rows along nearly straight lines. Those lines of volcanoes show that the magma rose along faults, undoubtedly the Basin and Range structures now creating the new fault block landscape of southern Idaho.

Hot Springs

Hot springs abound in southeastern Idaho, probably because it is within the region of most active Basin and Range faulting and the associated volcanic eruptions. Surface water sinks two or more miles along deep fractures in the fault zones, where it absorbs heat from the hot rocks at depth. Then the heated water rises to the surface for the same reason that it circulates through a home hot water heating system — because it is lighter than cold water.

Soda Springs was a famous landmark of the Oregon Trail. Emigrants eagerly anticipated enjoying its numerous hot springs, which produced carbonated hot water in a variety of flavors and to a distinctly anti-musical accompaniment of all sorts of rumbling, gurgling, and hissing noises. Individual springs had their distinctive temperatures, flavors, and uses.

Some of the waters were said to taste like stale beer, others more like stale lemonade. Bread dough mixed with the fizzing water rose as it baked to become loaves as light as those made with baking powder. Many of the springs built fanciful mounds of travertine around their vents as the emerging water precipitated part of its load of dissolved calcium carbonate. Idaho lost one of its most interesting sights when Soda Point Reservoir flooded most of that steaming and picturesque landscape of hot and flavorful curiosities.

An antique photograph of part of the Soda Springs basin as it looked more than a century ago. —U.S. Geological Survey photo by W.H. Jackson

Travertine deposits at Lava Hot Springs.

The state owns and operates Lava Hot Springs, ten miles west of Soda Springs, where water issues from a number of vents along the creek, most of them now enclosed to form pools. Large deposits of pale travertine, a form of limestone also known as calcareous tufa, mark the hot spring vents. Travertine deposits form as the cooling hot water precipitates its load of dissolved calcium carbonate.

The Clockwork Erupting Wells in Soda Springs

Some decades ago, a well drilled in Soda Springs tapped steaming hot water that fizzed vigorously as it unloaded its charge of dissolved carbon dioxide. The combined pressure of carbon dioxide and steam blew a towering column of hot water out of the hole. After the wild well was capped, it was fitted with a valve equipped with an automatic timer. At precisely regular intervals, the timer opens the valve to let the well spout its high column of steaming water. The valved and regulated well is truly an erupting well, no less genuine for being partly man made. The Soda Springs erupting well erupts with perfect regularity, and almost as spectacularly as Old Faithful. Over the years, the valved erupting well has built a gently sloping mound of travertine around its base.

367

The Soda Springs erupting well in full eruption.

Travertine Terraces

Mounds of travertine forming around hot springs tend to develop into flights of pretty pools enclosed within little dams. The dams form through an intricate chemical dance that involves water, carbon dioxide, and dissolved calcium carbonate.

When carbon dioxide gas dissolves in water, it forms carbonic acid, the same weak acid that makes soda water taste slightly sour. The weakly acidic solution of carbonic acid and water dissolves calcium carbonate. That is what makes hard water hard. The high pressures that prevail thousands of feet below the surface enable water there to dissolve much larger amounts of carbon dioxide, and therefore also of calcium carbonate, than it can hold under surface pressures.

That deep water, heavily charged with gas, emerges from the spring fizzing like soda pop as its load of dissolved carbon dioxide escapes in the lower pressures at the surface — the effect is exactly the same as that of taking the cap off a bottle of soda water. The loss of dissolved carbon dioxide makes the water less acid, less capable of holding dissolved calcium carbonate. So the calcium carbonate precipitates as travertine.

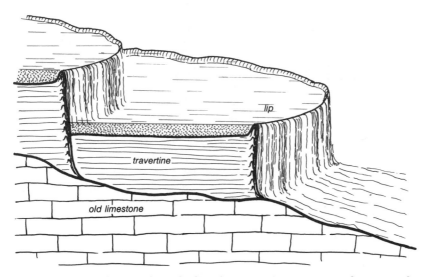

A cross-section showing how the lip of a travertine terrace tends to extend the surface area of water flowing over it, thus causing deposition of calcium carbonate on the lip.

The fizzing water precipitates calcium carbonate as it loses carbon dioxide gas, which must escape through the water surface. Any irregularity of flow that tends to increase the area of the water surface will promote the escape of carbon dioxide, thus localizing precipitation of calcium carbonate. So any irregularity in the mound of travertine around a hot spring tends to grow, eventually to become a miniature dam that encloses a little pool of water.

Phosphate Mines and Mills

A large open pit mine at Conda, a few miles north of Soda Springs, has produced phosphate rock from the Phosphoria formation ever since 1916. Several other phosphate mines north and east of Soda Springs boast impressively long histories of continuous operation. Phosphate mining is an important industry, certainly destined to become more so.

369

A phosphate fertilizer plant just north of Soda Springs. The black piles are coke used to fuel the plant.

We all need phosphate rock; it is an essential raw material for making fertilizer, certain kinds of detergents and explosives, and many industrial chemicals. Lesser amounts find their way into a wide variety of familiar uses that range from making the heads on matches to flavoring soft drinks.

Large deposits of phosphate rock in Florida have dominated the market for years, partly because the ore is good, mainly because it is close to major markets and to ocean shipping. The Rocky Mountain phosphate field in southeastern Idaho and nearby parts of Wyoming and Montana contains plenty of good ore, but suffers from the high shipping costs to major markets. That problem will certainly diminish as the reserves of good Florida ore dwindle and world demand for phosphate increases.

The primary phosphate mineral is apatite, calcium phosphate, the same mineral that forms bones and teeth. Phosphate rock can be used just as it comes from the mine as a slow acting and long lasting fertilizer. Processing plants treat much of the rock mined in south-eastern Idaho with sulfuric acid to make phosphoric acid, which is then reacted with more phosphate rock to make super phosphate, a much faster and more potent fertilizer. Other reactions combine treated phosphate rock with ammonia to make ammonium phosphate fertilizers.

The Phosphoria Formation

All of Idaho's phosphate rock comes from the Phosphoria formation, one of the least conspicuous and most valuable bodies of rock in the state. It was laid down on the floor of a shallow sea about 230

370

million years ago, during the Permian period, and extends into large areas of western Wyoming and southwestern Montana. It contains the world's largest known reserves of phosphate rock, as well as some oil shale that may also contain metals.

Phosphate rock, a rare kind of sedimentary rock, looks at first glance like black or dark brown limestone, but it is quite noticeably heftier than limestone, and distinctly harder. Weathered surfaces tend to develop a filmy coating of blue vivianite, an iron phosphate mineral. Stream rounded pebbles of phosphate rock commonly acquire a high polish that makes them look almost as though they had been polished in a tumbler.

Oil shale is also dark brown or black, but it is soft, flaky, and has a waxy appearance — not much to look at. But you can light oil shale with a match, then watch it burn with a yellow flame that smells bad, makes a lot of smoke, and leaves a truly excessive residue of ash. People need to be fairly desperate to decide to use oil shale for stove fuel.

The ash from most oil shales contains large amounts of such metals as copper, nickel, and chromium, among others. Several large mines in this country and in Europe produce from such metalliferous shales. We have not seen estimates of the metals content of the oil shales in the Phosphoria formation, but suspect they may be substantial.

It is indeed possible to roast genuine crude oil out of oil shale. Unfortunately, the reserves of oil shale in the Phosphoria formation are probably too small and to difficult to mine to qualify the formation as a significant source of petroleum. Nevertheless, they may well be large enough to provide the energy needed to recover whatever metals the rock may contain.

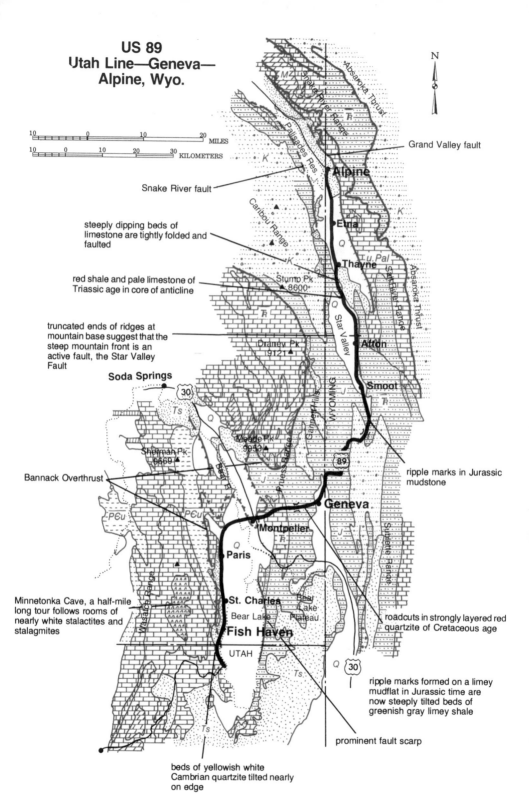

US 89
Utah Line—Geneva—
Alpine, Wyo.

10 0 10 20 MILES
10 0 10 20 30 KILOMETERS

N

Grand Valley fault

Snake River fault

steeply dipping beds of
limestone are tightly folded and
faulted

red shale and pale limestone of
Triassic age in core of anticline

truncated ends of ridges at
mountain base suggest that the
steep mountain front is an
active fault, the Star Valley
Fault

Soda Springs

Bannack Overthrust

Minnetonka Cave, a half-mile
long tour follows rooms of
nearly white stalactites and
stalagmites

Alpine

Etna

Thayne

Afton

Smoot

Geneva

Montpelier

Paris

St. Charles

Fish Haven

UTAH

ripple marks in Jurassic
mudstone

roadcuts in strongly layered red
quartzite of Cretaceous age

ripple marks formed on a limey
mudflat in Jurassic time are
now steeply tilted beds of
greenish gray limey shale

prominent fault scarp

beds of yellowish white
Cambrian quartzite tilted nearly
on edge

Absaroka Thrust

Palisades Res.

Caribou Range

Stump Pk.
8600

Draney Pk.
9121

Meade Pk.
9953

Sherman Pk.
9669

Bear Lake
Plateau

Bear Lake

Bear
Lake

Star Valley

Snake River Range

Salt River Range

Caribou Hills

Preuss Range

Wasatch Range

Sublette Range

WYOMING

Absaroka Thrust

382

U.S. 89:
Utah — Wyoming
44 miles

This beautiful drive follows the western shore of Bear Lake and crosses part of the southeastern Idaho overthrust belt. The road between the Utah line and Montpelier crosses deep deposits of unconsolidated sediments laid down in the floor of the Bear Lake Valley during the last few million years. Between Montpelier and the Wyoming line, the route follows picturesque Montpelier Canyon through excellent exposures of rocks in the overthrust belt to Geneva Summit, then down the eastern slope of the mountains through more of the overthrust belt.

The Overthrust Belt

Montpelier Canyon cuts right across the grain of the overthrust belt, and of the landscape. How can a river start flowing in a path so utterly contrary to the strong dictates of both bedrock and topography? In this case, the structure of the bedrock is so old, 70 to 80 million years, that the folds and faults were surely in place long before the modern stream began to flow. It seems more likely that the stream began to flow on soft valley-fill sediments, then cut through them into the older and harder rocks beneath.

Cliffs near Montpelier expose the pale gray Madison limestone. Rocks along the remainder of the route through the overthrust belt are much younger formations deposited during Permian, Triassic, and Jurassic time, between about 280 and 150 million years ago. All those rocks were crumpled into a series of folds, anticlinal arches and synclinal troughs, as they moved east into the overthrust belt, about 70 million years ago. Watch for the changes in the directions the sedimentary layers tilt.

As the overthrust belt formed, the thick Madison limestone was shoved up and over the younger formations to the east along the Bannock overthrust fault, one of the major bedrock structures in southeastern Idaho.

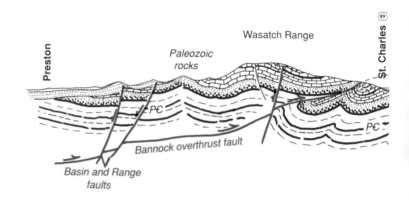

Section across the line of the highway and Bear Lake showing the Bannock

Bear Lake Valley

Between Montpelier and the Utah line, the route follows the north end and western side of a large lowland, part of the Bear Lake Valley. Throughout that part of the route, the highway crosses soft sedimentary deposits accumulated within the last million or so years.

Bedrock in the hills west of the highway between the Utah line and Ovid is mostly Paleozoic sedimentary formations tightly scrunched up in the overthrust belt. Rocks immediately west of the highway, those you can actually see from the road, include younger formations deposited during the early part of Mesozoic time, also scrunched. The older rocks in the hills farther west rode up over them on the Bannock overthrust fault.

Bear Lake

Bear Lake Valley appears to be part of the Basin and Range, a depression formed as a block of the earth's crust dropped along faults. Notice the remarkably straight mountain wall along the eastern side of the lake. It hardly looks eroded. A mountain wall that straight and steep strongly suggests an active fault scarp rising more rapidly than the slow processes of erosion can carve it. The eastern side of the lake near that wall is also its deepest part, about 210 feet.

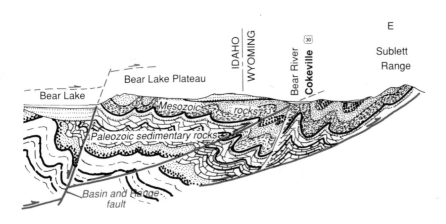

overthrust fault and the much younger Basin and Range faults that cut it.

Watch for the old shoreline about 33 feet above the present high water level of the lake. The lake probably stood at that level during the ice ages, at the same time the Great Salt Lake was expanded into Lake Bonneville. When the lake stood at that ice age level, the Bear River flowed through it, instead of passing across the floor of the Bear Lake Valley eight miles to the north.

The Bear River-Bear Lake Irrigation Project, completed in 1918, restores part of that ice age connection between river and lake. A canal near Dingle diverts the spring flood into Bear Lake, which stores the water until irrigation demand rises during the summer. Then another canal returns the stored flood water to the Bear River just west of Montpelier. Seasonal fluctuation of the lake level averages between three and four feet.

Bear Lake has a peculiarly vivid greenish color that most people find extraordinarily beautiful, almost haunting. Some authorities attribute that distinctive color to a high content of dissolved carbonates in its water, but the chemical analyses we have seen don't reveal anything so unusual. Bear Lake water is, after all, perfectly fit for irrigation. We think it more likely that green plants growing on the shallow floor of the lake combine with the reflected color of the sky to give Bear Lake its marvelous color.

Wandering Rivers

The Bear River leaves the Bear Lake Valley, flows north and west through a narrow canyon carved into hard bedrock, crosses the Gem Valley, flows through another canyon sliced into hard bedrock, then enters the broad Salt Lake Valley. That is an amazingly adventurous course for such a modest stream, quite a tour of the countryside. The equally vagrant Portneuf River follows a circuitous path through a corner of the Gem Valley, past McCammon, and then into the Snake River.

We think the Bear and Portneuf rivers, like the other major streams in the region, began their careers by overflowing from one formerly undrained desert valley to another after the climate became wetter at the beginning of Pleistocene time, about 2.5 million years ago. Nevertheless, it seems extremely peculiar that both the Bear and Portneuf rivers go through the Gem Valley — that can hardly be the result of overflowing lakes. It is equally peculiar that the Portneuf Valley is much too big for its little river; it is about the right size to accomodate the much larger Bear River.

The wandering Bear and Portneuf Rivers, and the valleys they cross.

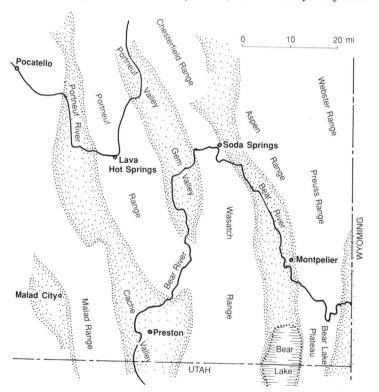

We suggest that the Bear River may originally have flowed down the present valley of the Portneuf River. Then the enormous Portneuf lava flow that probably started in the Gem Valley and flowed 40 miles down the Portneuf River Valley diverted the original lower part of the Bear River into the Salt Lake Valley. After that diversion, the former tributaries of the Bear River continued to empty into its old valley to form the modern Portneuf River.

Active faults, as well as lava flows, complicate the landscape of this part of Idaho. All the mountains the Bear and Portneuf Rivers slice through and all the valleys they cross are actively moving Basin and Range fault blocks. The mountains have risen and the basins dropped while the modern rivers eroded their valleys. Meanwhile, the streams kept pace with those movements by eroding sediment from their canyons and depositing it in the valleys to maintain their graded courses across an unevenly moving landscape. Except in its failure to drain Bear Lake, the Bear River has successfully coped with the tectonic movements of the ranges and valleys it crosses.

Salt

One of Idaho's earliest industries produced salt from a number of localities along Stump and Crow creeks, just inside the state line in the area west of Auburn and Fairfield, Wyoming. The Lander Cutoff part of the Oregon Trail follows Stump Creek, so these salt deposits that now seem so remote were conveniently on one of the main routes of travel in early Idaho.

Salt production began in 1866 with an operation that involved wooden pipes carrying water from a saline spring to large metal pans, where it was evaporated over open fires to concentrate the salt. The crystallized salt, sold in bags, was widely celebrated for its purity. In the early days, it sold at the evaporating plant for about $30 a ton, and was shipped in wagons to markets throughout much of Idaho and western Montana. Several other saline springs in eastern Idaho were soon put into commercial salt production.

Total production from all the springs rose to a maximum of approximately 800 tons a year in 1884. Then output dropped rapidly as railroads brought the sales region into easy shipping distance of underground mines in Utah that could produce a ton of rock salt for $3.50. Annual production from all the salt springs of eastern Idaho was down to a few tons by the turn of the century.

The Idaho salt industry revived for a while after 1902, when solid rock salt was discovered a few feet beneath one of the springs. But the revival was brief because the rock salt contains so much red clay that it is suitable mostly for cattle, who don't seem particular about the color of their salt. The springs produced extremely pure salt from those highly impure deposits because the water dissolved only the salt, leaving the clay behind.

All the salt is in the Preuss formation, which was deposited during Jurassic time, about 150 to 200 million years ago. In most areas the Preuss formation consists of sandstone and shale deposited in shallow sea water. In this part of Idaho it also contains layers of salt, presumably laid down in an evaporating coastal lagoon.

Minnetonka Cave

Limestone caverns form as rainwater made slightly acidic by reaction with carbon dioxide in the atmosphere dissolves passages in the rock. Then water percolating through the limestone drips into those open passages, where it loses part of its dissolved carbon dioxide. As carbon dioxide escapes from the water, calcium carbonate precipitates to become the fantastic travertine formations: stalactites that hang from the roof, stalagmites that rise like ornate stumps from the floor, and all sorts of draperies that form where water trickles along the wall, instead of dripping from cave ceiling to floor.

The Works Progress Administration of the depression years built trails and installed steps to develop Minnetonka Cave for visitor access. The cave has been open to the public since 1937, during the summer. The tour, which takes about an hour, is a rare treat. This is one of the few large limestone caverns in Idaho, a state far more notable for its many lava caves in basalt flows — a totally different sort of thing.

If You Want To Read More

The geologic literature of any state tends to be difficult and frustrating. Idaho is no exception. Most of the material consists of papers dealing with specific topics in language so filled with technical jargon you need a dictionary to read them. Furthermore, you can find most them only in university libraries. Some public libraries do keep the publications of the Idaho Bureau of Mines and Geology in their Idaho collections.

The easiest and best place to start is with a copy of the state geologic map, a brightly colorful sheet beautiful enough to deserve hanging on the wall. You can buy it from the Idaho Geological Survey (Bureau of Mines and Geology) at the university in Moscow. If your friends are not interested in rocks, tell them to enjoy it as abstract art.

The libraries of the University of Idaho in Moscow, Idaho State University in Pocatello, and Boise State University in Boise all have reasonably good collections of material dealing with Idaho geology, including the references below. We cite them in their proper bibliographic form to make it easier for libraries to help enable you to get started reading about many of the major topics in Idaho geology. You can find further references in the bibliographies in each paper.

Alt, D., Sears, J.W., and Hyndman, D.W., 1988, *Terrestrial maria: The origins of large basalt plateaus, hotspot tracks, and spreading ridges*: Journal of Geology, v. 96, p. 647-662.
(This article shows that a large meteorite struck southeastern Oregon about 17 million years ago, starting the Columbia Plateau, Snake River Plain, and Basin and Range.)

Bond, John G., 1978, Geologic Map of Idaho: Idaho Bureau of Mines and Geology, Moscow, Scale 1:500,000
(This is the beautiful state map we mentioned above, the single most essential reference on the geology of Idaho).

Bonnichsen, Bill, and Breckenridge, R.M., 1982, *Cenozoic Geology of Idaho:* Idaho Bureau of Mines and Geology Bulletin 26, 725 p.
(This thick book is full of important papers on most aspects of Idaho geology of the last 65 million years, an indispensable reference. It provides excellent information about Challis volcanic rocks, the Columbia Plateau, Snake River Plain, fossils, ice age glaciation, and drainage of Lake Bonneville.)

Hamilton, W.B., 1963, *Metamorphism in the Riggins region, western Idaho:* U.S. Geological Survey Professional Paper 436, 95 p.
(A long paper that described the rocks of the Seven Devils and Riggins complexes long before anyone realized that they are former oceanic islands.)

McIntyre, D.H. (editor), 1985, *Symposium on the geology of mineral deposits of the Challis 1 x 2 quadrangle, Idaho*: U.S. Geological Survey Bulletin 1658.
(This collection of articles deals with the ore deposits of a large part of east central Idaho. The references point to other papers dealing with other aspects of Idaho ore deposits.)

Smith, r.b. and Christiansen, R.L., 1980, *Yellowstone Park as a window on the Earth's interior*: Scientific American, v. 242, p 104.
(This authoritative article on the Yellowstone volcano is available in most public libraries.)

Smith, R.B. and others, 1989, *What's moving at Yellowstone?*: Transactions of the American Geophysical Union, EOS, v. 70, no. 8, p. 113-125.
(A more technical discussion of the Yellowstone volcano for the more determined reader.)

Smith, R.b. and eaton, G.P. (editors), *Cenozoic tectonics and regional geophysics of the western Cordillera*: Geological Society of America Memoir 152
(A collection of articles on formation of the Rocky Mountains.)

Swanson, D.A., Wright, T.L., Hooper, P.R., and Bentley, R.D., 1979, *Revisions in stratigraphic nomenclature of the Columbia River basalt group*:U.S. Geological Survey Bulletin 1457-G, G1 - G59.
(Here you will find excellent descriptions of the basalt flows that make the Columbia Plateau.)

Waitt, Richard B., 1985, *Case for periodic collosal jokulhlaups from Pleistocene Glacial Lake Missoula*: Geological Society of America Bulletin, p. 1271-1286.
(This discussion of the drainages of Glacial Lake Missoula also contains references to most of the other important literature on the subject.)

Glossary

Agate: A specimen of the mineral chalcedony that is pretty enough to cut and polish as a semi-precious stone.

Andesite: A common volcanic rock intermediate in composition between basalt and rhyolite. Comes in shades of gray or brown.

Anticline: A fold that arches up.

Ash: Fine shreds of lava blown from a volcano.

Ash Flow: A deposit of volcanic ash, typically rhyolite, laid down from a ground-hugging cloud of red hot steam and ash.

Asthenosphere: A zone of partially molten rock within the earth's mantle. The top of the asthenosphere is at a depth of about 40 to 100 or more miles, and it extends to a depth of about 80 to 150 or more miles. Lithospheric plates move by sliding on the asthenosphere.

Augite: A black mineral common in dark igneous and metamorphic rocks.

Basalt: A black volcanic rock rich in iron, calcium, and magnesium and composed primarily of augite and plagioclase feldspar. Basalt is the commonest volcanic rock.

Basement Rock: The complex assortment of gneiss, schist, and granite that composes the continental crust.

Basin and Range: A geologic province of western North America in which faults that trend generally north are breaking the earth's crust into blocks, and pulling them away from each other.

Bedrock: Rock firmly in place. Untransported.

Belt Rock: One of the so-called Belt formations of sedimentary rocks, which were laid down in the northern Rocky Mountains during the latter part of Precambrian time, between about 1500 and 800 million years ago.

Biotite: Common black mica. It occurs as black flakes in many granites and metamorphic rocks.

Bomb: A glob of lava, typically basalt, blown from a volcano by escaping steam. Most bombs are streamlined by their molten passage through the air.

Calcite: The mineral form of calcium carbonate. Calcite is the main substance of limestone, marble, sea shells, pearls, and many other familiar objects.

Caldera: A collapse crater formed as large amounts of lava erupt from a magma chamber below the surface.

Chalcedony: A form of quartz composed of very finely fibrous crystals. Chalcedony is common in petrified wood, and fills open spaces in many volcanic rocks. If colorful, it may be called agate.

Cinder Cone: A volcano composed mostly of bits of basalt blown from the vent by escaping steam.

Continental Crust: The complex assortment of light igneous and metamorphic rocks, basement rocks, that lies beneath most continental areas.

Crust: The outer skin of the earth, either continental crust or oceanic crust.

Diabase: An igneous rock similar to basalt except in being composed of larger mineral grains. Diabase typically consists of pale crystals of plagioclase feldspar scattered through a matrix of black augite.

Dike: A body of igneous rock formed when molten magma injected a fracture.

Diorite: A coarsely crystalline igneous rock that consists mostly of feldspar, along with black hornblende and mica. Diorite resembles granite except in being darker and containing little or no quartz.

Dolomite: The form of calcium-magnesium carbonate. It forms a sedimentary rock of the same name that looks much like limestone.

Erratic: A piece of rock that has been transported from its original home. Glacially transported erratics are very common in northern Idaho.

Fault: A fracture in the earth's crust along which the rocks on one side have moved past those on the other.

Feldspar : A family of minerals composed of aluminum, silica, and either sodium and calcium in plagioclase or potassium and sodium in orthoclase. Feldspars are the most abundant minerals in the earth's crust.

Flood Basalt: As the expression suggests, a basalt lava flow of overwhelming size.

Formation: Any body of whatever kind of rock that can be recognized from one place to another.

Garnet: A family of silicate minerals common in many kinds of igneous and metamorphic rocks, as well as in placer deposits. Most garnets are red.

Gneiss: A metamorphic rock composed mostly of feldspar that displays distinct layering. Most gneisses resemble granite except in having a generally streaky appearance.

Granite: A common igneous rock composed mostly of orthoclase and plagioclase feldspar along with quartz and either mica or hornblende, all in crystals large enough to distinguish with the unaided eye. Granite and rhyolite both have the same composition, but differ in crystal size.

Hornblende: A glossy black silicate mineral that crystallizes into long prisms. Hornblende is common in light colored igneous or metamorphic rocks such as granite and gneiss.

Hotspot: A volcanic center that has no apparent association with plate boundaries; the Yellowstone volcano, for example.

Kaolinite: One of many kinds of clay minerals that commonly forms in the soils of humid regions. Kaolinite is useful in making fire brick, porcelain, and many kinds of specialty ceramics.

Lava: Molten rock that has erupted through a volcano

Lava Tube: A tunnel within a basalt lava flow that formed as still molten lava ran out from under an already hardened outer crust.

Lithosphere: The relatively rigid outer rind of the earth that includes the crust and the upper part of the mantle down to the top of the asthenosphere.

Limestone: A common sedimentary rock composed mostly of the mineral calcite, calcium carbonate.

Lode: Ore deposit in bedrock, as opposed to placer deposits in stream or beach gravels and sands.

Magma: Molten rock. Technically, the term magma applies as long as the molten rock is below the surface. If it erupts through a volcano, the molten rock becomes lava.

Mantle: The part of the earth between the crust and the core, most of the planet. Rocks in the mantle are varieties of peridotite.

Mesozoic: The interval of time that began about 240 million years and ended about 65 million years ago. It includes the Triassic, Jurassic, and Cretaceous periods, from oldest to youngest.

Metamorphic Rock: A rock formed through recrystallization of an older rock at high temperature.

Metamorphism: Recrystallization of an older rock at high temperature. Metamorphism commonly, but not necessarily, occurs deep in the earth's crust under conditions of high pressure.

Mica: A group of silicate minerals in which the molecular structure is layered. Most micas reflect that internal structure by splitting easily into thin flakes. The common kinds of mica are biotite, which is black, and muscovite, which is white.

Moraine: A deposit of glacial till.

Mother Lode: The bedrock source of a mineral, such as gold, that occurs in a placer deposit.

Muscovite: The common variety of white mica. Look for glistening flakes of muscovite in many granites and metamorphic rocks.

Mylonite: A metamorphic rock that was strongly sheared while hot enough to flow, instead of breaking. Faults turn into mylonites as they penetrate into the deeper levels of the crust.

Obsidian: Volcanic glass, a magma that hardened into solid rock without crystallizing. Most obsidian has the same composition as rhyolite or granite.

Oceanic Crust: The complex of basalt lava flows and related rocks that forms at oceanic ridges, and lies beneath the oceans.

Oceanic Ridge: A plate boundary at which two plates pull away from each other as basalt erupting into the gap between them forms new oceanic crust.

Opal: A mineral that consists of uncrystallized silica. Shapeless masses of dull yellow or green opal commonly fill fractures and other openings in volcanic rocks. They form as water percolates through the rock, long after it has cooled.

Olivine: An iron and magnesium silicate mineral common in black igneous rocks such as basalt. Olivine typically forms small crystals that look like bits of greenish glass. It weathers easily into rusty specks.

Ore: Any body of rock that can be profitably mined.

Orthoclase: A variety of feldspar that contains potassium. Orthoclase, with plagioclase, is the major mineral in granites, as well as in most kinds of gneiss and schist. It typically crystallizes into blocky grains that are either pink, beige, or white, never greenish as in plagioclase.

Outcrop: A natural exposure of bedrock.

Outwash: Sediment deposited from glacial meltwater.

Overthrust Belt: A broad zone in which numerous overthrust faults have stacked slabs of rock on each other in an overlapping pattern like that of shingles on a roof.

Overthrust Fault: A very gently inclined fault along which a slab of rock rode over the rocks beneath. Most overthrust faults place older rocks on top of younger ones.

Paleozoic: The interval of time that ran from about 570 million years to about 240 million years ago. It includes, from oldest to youngest, the Cambrian, Ordovician, Silurian, Devonian, Mississippian, Pennsylvanian, and Permian periods.

Pegmatite: An igneous rock composed of extremely large crystals.

Peridotite: A black igneous rock composed primarily of the minerals augite and olivine. Peridotite comprises the mantle, and is therefore the most abundant rock in the earth.

Phosphate Rock: Rock composed largely of the phosphate mineral apatite. Phosphate rock is typically brown or black, and distinctly denser than most common rocks.

Placer: A concentration of heavy minerals in stream or beach sands. Minerals commonly found in placers include magnetite, garnet, gold, and zircon.

Plagioclase: Variety of feldspar that contains sodium and calcium in variable proportions. Plagioclase typically crystallizes into blocky grains that are white or greenish white, never pink as orthoclase.

Precambrian The long span of time from the formation of the earth to the beginning of Cambrian time, about 570 million years ago.

Pressure Ridge: A raised ridge on a lava flow that formed as slabs of hardened basalt riding on molten lava beneath jammed together. A long crack follows the crest of most pressure ridges.

Plate: A segment of the earth's lithosphere.

Pyrite: An iron sulfide mineral common in many kinds of rocks. Pyrite typically crystallizes into cubes the color of brass that have a metallic appearance. Many people call it "fool's gold."

Quartz: The common mineral form of silicon dioxide, the most widely distributed mineral on the earth's surface. Quartz occurs in a wide variety of forms, most commonly in grains that look like shapeless bits of glass.

Radiocarbon: A method of determining the age of something by measuring its content of radioactive carbon-14. Works only on material less than 40,000 years old.

Resurgent Caldera A type of volcano that typically produces enormous volumes of rhyolite lava in widely spaced eruptions.

Rhyolite: A very pale volcanic rock that has the same composition as granite.

Schist: A metamorphic rock that contains enough mica to confer a flaky quality.

Sediment: Material such as sand, mud, or gravel carried by running water, waves, or wind.

Sedimentary Rock: A deposit of sediment hardened into rock. Most sedimentary rocks are layered, stratified.

Shield Volcano: A volcano composed mostly of basalt lava flows. Shield volcanoes are typically broad and relatively flat, shaped approximately like a vanilla wafer.

Silica: Silicon dioxide, the most abundant substance in the earth's crust and mantle.

Silicate: A mineral based on silica. The silicate minerals form most of the earth's crust and mantle.

Sill: A layer of igneous rock sandwiched between layers of sedimentary rock.

Sillimanite: An aluminum silicate mineral that forms in high temperature metamorphic rocks, like schist and gneiss. It typically occurs as minute fibers that become visible only with a strong magnifying glass.

Silt: Sediment finer than sand, coarser than clay. Recognize siltstone by dragging your fingernail across the rock. It feels about as rough as a fine emery board.

Slate: Slightly metamorphosed shale or mudstone that splits into thin slabs along a direction that cuts across the original sedimentary layering.

Spatter Cone: A small volcanic cone formed as escaping steam coughs blobs of magma out of a lava flow.

Syncline: A fold that arches down, like a trough.

Thrust Fault: A type of fault in which the rocks above the fracture surface moved up, onto those below. Thrust faulting shortens the earth's crust.

Thrust Plate: A slab of rock, typically thousands of feet thick, that moved above a thrust fault.

Till: Sediment deposited directly from glacial ice. Most till consists of a disorderly mixture of all sizes of rocks suspended in a muddy matrix.

Trench: A long and very deep trough in the ocean floor that forms above a sinking lithospheric plate.

Transform Fault: A boundary along which two plates slide past each other.

Tuff: A deposit of pale volcanic ash.

Vein: A body of rock that fills a fracture in older rock.

Index

We encourage you to patronize your local bookstore. Most stores will order any title they do not stock. You may also order directly from Mountain Press, using the order form provided below or by calling our toll-free, 24-hour number and using your VISA, MasterCard, Discover or American Express.

Some geology titles of interest:

____ROADSIDE GEOLOGY OF ALASKA	18.00
____ROADSIDE GEOLOGY OF ARIZONA	18.00
____ROADSIDE GEOLOGY OF COLORADO, 2nd Edition	20.00
____ROADSIDE GEOLOGY OF HAWAII	20.00
____ROADSIDE GEOLOGY OF IDAHO	20.00
____ROADSIDE GEOLOGY OF INDIANA	18.00
____ROADSIDE GEOLOGY OF MAINE	18.00
____ROADSIDE GEOLOGY OF MASSACHUSETTS	20.00
____ROADSIDE GEOLOGY OF MONTANA	20.00
____ROADSIDE GEOLOGY OF NEBRASKA	18.00
____ROADSIDE GEOLOGY OF NEW MEXICO	18.00
____ROADSIDE GEOLOGY OF NEW YORK	20.00
____ROADSIDE GEOLOGY OF NORTHERN and CENTRAL CALIFORNIA	20.00
____ROADSIDE GEOLOGY OF OREGON	16.00
____ROADSIDE GEOLOGY OF PENNSYLVANIA	20.00
____ROADSIDE GEOLOGY OF SOUTH DAKOTA	20.00
____ROADSIDE GEOLOGY OF TEXAS	20.00
____ROADSIDE GEOLOGY OF UTAH	20.00
____ROADSIDE GEOLOGY OF VERMONT & NEW HAMPSHIRE	14.00
____ROADSIDE GEOLOGY OF VIRGINIA	16.00
____ROADSIDE GEOLOGY OF WASHINGTON	18.00
____ROADSIDE GEOLOGY OF WISCONSIN	20.00
____ROADSIDE GEOLOGY OF WYOMING	18.00
____ROADSIDE GEOLOGY OF THE YELLOWSTONE COUNTRY	12.00
____COLORADO ROCKHOUNDING	20.00
____FIRE MOUNTAINS OF THE WEST	18.00
____GEOLOGY OF THE LAKE SUPERIOR REGION	22.00
____GEOLOGY OF THE LEWIS AND CLARK TRAIL IN NORTH DAKOTA	18.00
____GEOLOGY UNDERFOOT IN CENTRAL NEVADA	16.00
____GEOLOGY UNDERFOOT IN DEATH VALLEY AND OWENS VALLEY	16.00
____GEOLOGY UNDERFOOT IN ILLINOIS	15.00
____GEOLOGY UNDERFOOT IN SOUTHERN CALIFORNIA	14.00
____GLACIAL LAKE MISSOULA AND ITS HUMONGOUS FLOODS	15.00
____NORTHWEST EXPOSURES	24.00

Please include $3.00 per order to cover postage and handling.

Send the books marked above. I enclose $_____

Name _____

Address _____

City/State/Zip _____

☐ Payment enclosed (check or money order in U.S. funds)

Bill my: ☐ VISA ☐ MasterCard ☐ Discover ☐ American Express

Card No. _____ Expiration Date:_____

Signature _____

MOUNTAIN PRESS PUBLISHING COMPANY
P.O. Box 2399 • Missoula, MT 59806 • Order Toll-Free 1-800-234-5308
E-mail: info@mtnpress.com • Web: www.mountain-press.com